Graphing Calculator Instruction Guide

To Accompany

Calculus Concepts: An Informal Approach to the Mathematics of Change

LaTorre/Kenelly/Fetta/Carpenter/Harris

First Edition

Iris Brann Fetta
Clemson University

Houghton Mifflin Company   Boston   New York

Editor in Chief:	Charles Hartford
Associate Editor:	Elaine Page
Marketing Managers:	Ros Kane, Sara Whittern
Manufacturing Coordinator:	Florence Cadron

Printed in the U.S.A.

ISBN: 0-669-39864-0

56789-HS-01 00 99

PREFACE

This *Guide* uses actual examples and applications given in *Calculus Concepts: An Informal Approach to the Mathematics of Change*. Wherever a 📖 appears in your text, a new technology technique is illustrated in this *Guide*. The icon is your clue to refer to the particular section of this *Graphing Calculator Instruction Guide* that discusses the text section for your specific calculator. When there is no reference to a certain section in the text, either there is no new procedure to learn or the necessary techniques have been covered in an earlier section of this *Guide*. However, there are discussions in this *Guide* that refer to sections of your text not marked with a 📖. Refer to the table of contents of this *Guide* often.

The use of technology is an integral part of your study of calculus using *Calculus Concepts*. Even though the text requires technology, it does not demand any particular technology. Any graphing calculator or computer algebra system that has the functionality indicated in this *Guide* will suffice. The materials contained herein provide instruction for using the TI-82, TI-83, TI-85, or TI-86 in your course. A supplement to this *Guide*, covering exactly the same material as that discussed for the other calculators, is available for the HP 48G/GX.

This *Guide* is broken into two parts, with each part containing all the instruction for two particular calculator models. Within each part, the discussions are ordered to match the organization of the text chapters. You should refer to these materials for explanations of how to use your calculator with *Calculus Concepts* as you go through each section of the text.

Throughout this *Guide*, the following notation conventions will be used to help you recognize various commands and keystrokes:

- Main keyboard keys are enclosed in rectangular boxes (for example, ENTER) except for certain numeric keys, English alphabet letters, and the decimal point.

- The second function of a key is listed in parentheses after the main keyboard keystrokes used to activate the second function (for example, 2nd LN (e^x)).

- The alpha function of a key is listed in parentheses after the main keyboard keystrokes used to activate the alpha function (for example, ALPHA SIN (E)).

- Function keys and menu items are indicated by the main keyboard key followed by the keystroke sequence necessary to access the item and the name of the item (for example, STAT 1 (Edit)).

The calculator code for programs referenced in these materials are listed in a separate *Appendix* for each of the two parts of the manual.

This *Guide* does not replace your calculator's instruction manual. You should refer to that manual to learn about the basic operation and use of your calculator.

Each of the calculators discussed in this *Guide* has special functionality in certain areas that the other calculators do not. The instructions in this *Guide* cover only those techniques that can be used on all the calculators because many classes using *Calculus Concepts* are taught to students in multi-calculator classrooms.

The author team of *Calculus Concepts* gives special thanks to these calculator manufacturers: Texas Instruments, Sharp, Hewlett-Packard, and Casio, for their assistance during the preparation and field testing of the text manuscript and *Guide*.

Any comments or suggestions concerning this *Guide* can be directed to the publisher or to

Iris B. Fetta

Clemson University
Mathematical Sciences
Clemson, SC 29634-1907

ibbrh@clemson.edu

Graphing Calculator Instruction Guide

for

Calculus Concepts:
An Informal Approach to the Mathematics of Change

CONTENTS

Detailed contents begin on page vi.

HOW TO USE THIS GUIDE

Use the following table to locate the part of this *Guide* in which your calculator model is discussed:

Part	Calculator Model
A	Texas Instruments TI-82 and Texas Instruments TI-83
B	Texas Instruments TI-85 and Texas Instruments TI-86

The section references in the following detailed table of contents should be read as follows:

> Section $x.y.z$ of this *Guide* refers to discussion in Chapter x, Section y of the *Calculus Concepts* text. The discussion is the zth topic in the *Guide* that refers to that particular chapter and section.

For example, the detailed contents indicate that there are 3 different sections in the *Guide* (1.2.1 through 1.2.3) pertaining to Section 1.2 of *Calculus Concepts*. The discussion in Section 1.2.2 of this *Guide* is the second technology discussion pertaining to Chapter 1, Section 1.2 of your text.

In Part *A*, all instruction is for both the TI-82 and TI-83 (or in part *B*, for the TI-85 and TI-86) unless it is marked for a particular calculator. When the common discussion resumes for both calculators, it is so indicated.

DETAILED CONTENTS

PART A

TEXAS INSTRUMENTS TI-82/TI-83 GRAPHICS CALCULATORS

Setup

When using this *Guide*, you should always, unless instructed otherwise, use the calculator setup specified below:

FOR THE TI-82 Before you begin, check the TI-82's basic setup by pressing MODE. Choose the settings shown in Figure 1. Check the statistical setup by pressing STAT ▶ (CALC) 3 (SetUp). Choose the settings shown in Figure 2. Check the window format by pressing ENTER ▶ (FORMAT), and choose the settings shown in Figure 3.

- If you do not have the darkened choices shown in each of the figures below, use the arrow keys to move the blinking cursor over the setting you want to choose and press ENTER.

- Press 2nd MODE (QUIT) to return to the home screen.

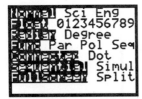

TI-82 Basic Setup

FIGURE 1

TI-82 Statistical Setup

FIGURE 2

TI-82 Window Setup

FIGURE 3

FOR THE TI-83 Before you begin, check the TI-83's basic setup by pressing MODE. Choose the settings shown in Figure 4. Specify the statistical setup by pressing 2nd MODE (QUIT) STAT 5 (SetUpEditor) followed by 2nd 1 , 2nd 2 , 2nd 3 , 2nd 4 , 2nd 5 , 2nd 6 . Choose the settings shown in Figure 2. Press ENTER to view the screen in Figure 5. Also, check the window format by pressing 2nd ZOOM (FORMAT), and choose the settings shown in Figure 6.

- If you do not have the darkened choices shown in each of the figures below, use the arrow keys to move the blinking cursor over the setting you want to choose and press ENTER .

- Press 2nd MODE (QUIT) to return to the home screen.

TI-83 Basic Setup
FIGURE 4

TI-83 Statistical Setup
FIGURE 5

TI-83 Window Setup
FIGURE 6

Basic Operation

You should be familiar with the basic operation of your calculator. With calculator in hand, go through each of the following.

1. **CALCULATING** You can type in lengthy expressions; just make sure that you use parentheses when you are not sure of the calculator's order of operations. As a general rule, numerators and denominators of fractions and powers consisting of more than one term should be enclosed in parentheses.

Evaluate $\dfrac{1}{4*15+\dfrac{895}{7}}$.
 	
 Evaluate $\dfrac{(-3)^4-5}{8+1.456}$. (Use $\boxed{(-)}$ for the negative symbol and $\boxed{-}$ for the subtraction sign.)
	`1/(4*15+895/7)`
	 `          .0053231939`
	 `((-3)^4-5)/(8+1.`
	 `456)`
	 `          8.037225042`

Evaluate $e^3*0.027$ and $e^{3*0.027}$.
 	
 The calculator will assume you mean the first expression unless you use parentheses around the two values in the exponent. (It is not necessary to type in the 0 before the decimal point.)
	`e^3*.027`
	 `          .5423094969`
	 `e^(3*.027)`
	 `          1.084370897`

2. **USING THE *ANS* MEMORY** Instead of again typing an expression that was evaluated immediately prior, use the answer memory by pressing 2nd (-) (ANS).

Calculate $\left(\dfrac{1}{4*15+\dfrac{895}{7}}\right)^{-1}$ using this nice shortcut.
 	
 If you wish to clear the home screen, press CLEAR
	`895/7`
	 `          127.8571429`
	 `1/(4*15+Ans)`
	 `          .0053231939`
	 `Ans⁻¹`
	 `          187.8571429`

3. ANSWER DISPLAY   When the denominator of a fraction has no more than three digits, your calculator can provide the answer in fraction form. When an answer is very large or very small, the calculator displays the result in scientific notation.

The "to a fraction" key is obtained by pressing MATH 1 (▸Frac).	`2/5+1/3`
	 `          .7333333333`
	 `Ans▸Frac`
	 `                11/15`
	 `.3875▸Frac`
	 `                31/80`

The calculator's symbol for "times 10^{12}" is **E**12. Thus, 7.945**E**12 means 7,945,000,000,000.
 	
 The result 1.4675**E**–6 means $1.4675*10^{-6}$, the scientific notation expression for 0.0000014675.
	`5600000000000+23`
	 `45000000000`
	 `            7.945E12`
	 `.00025*.00587`
	 `            1.4675E-6`

4. STORING VALUES   Sometimes it is beneficial to store numbers or expressions for later recall. To store a number, type the number on the display and press STO▸ ALPHA, type the letter in which you wish to store the value, and then press ENTER. To join several short commands together, use 2nd . (:).

Store 5 in A and 3 in B, and then calculate $4A - 2B$.
 	
 To recall a value stored in a variable, use ALPHA to type the letter in which the expression or value is stored and then press ENTER. The value stays stored until you change it.
	`5→A:3→B`
	 `                    3`
	 `4A-2B`
	 `                   14`
	 `A`
	 `                    5`

5. ERROR MESSAGES   When your input is incorrect, an error message is displayed.

If you have more than one command on a line without the commands separated by a colon (:), an error message results when you press ENTER.	`2→AX+2→B▮`

TI-82 Choose 1 (Goto) to position the cursor to the place the error occurred so that you can correct the mistake or choose 2 (Quit) to begin a new line on the home screen.	`ERR:SYNTAX`
	 `▮▐Goto`
	 `2:Quit`

TI-83 Choose 2 (Goto) to position the cursor to the place the error occurred so that you can correct the mistake or choose 3 (Quit) to begin a new line on the home screen.	`ERR:SYNTAX`
	 `▮▐Quit`
	 `2:Goto`

Chapter 1 Ingredients of Change:
Functions and Linear Models

📖 1.1 Fundamentals of Modeling

There are many uses for a function that is entered in the graphing list. Graphing the function in an appropriate viewing window is one of these. Because you must enter all functions on one line (that is, you cannot write fractions and exponents the same way you do on paper) it is very important to have a good understanding of the calculator's order of operations and to use parentheses whenever they are needed.

1.1.1 ENTERING AN EQUATION IN THE GRAPHING LIST
Press $\boxed{Y=}$ to access the graphing list. The graphing list contains space for 10 equations, and the output variables are called by the names Y1, Y2, ..., and Yo. When you intend to graph an equation you enter in the list, you must use X as the input variable.

If there are any previously entered equations that you will no longer use, delete them from the graphing list.	Position the cursor on the line with the equation, and press $\boxed{\text{CLEAR}}$.
TI 82 Suppose you want to graph $A = 1000(1 + 0.05)^t$. For convenience, we use the first, or Y1 , location in the list. We intend to graph this equation, so enter the right hand side as 1000(1 + 0.05)^X. (Type X by pressing $\boxed{X,T,\theta}$, not the times sign $\boxed{\times}$.)	```Y1⊟1000(1+.05)^X Y2= Y3= Y4= Y5= Y6= Y7=```
TI 83 Suppose you want to graph $A = 1000(1 + 0.05)^t$. For convenience, we use the first, or Y1 , location in the list. We intend to graph this equation, so enter the right hand side as 1000(1 + 0.05)^X. (Type X by pressing $\boxed{X,T,\theta,n}$, not the times sign $\boxed{\times}$.) Plot1, Plot2, and Plot3 at the top of the Y= list should not be darkened. If any of them are, use ◣◤ until you are on the darkened plot name. Press $\boxed{\text{ENTER}}$ to make the name(s) not dark.	```Plot1 Plot2 Plot3 \Y1⊟1000(1+.05)^ X \Y2= \Y3= \Y4= \Y5= \Y6=```

1.1.2 DRAWING A GRAPH
If you have not already done so, enter the equation in the Y= list using X as the input variable before drawing a graph. We now draw the graph of $y = 1000(1 + 0.05)^x$.

Press $\boxed{\text{ZOOM}}$ $\boxed{4}$ (ZDecimal). Notice that the graphics screen is blank.	```ZOOM MEMORY 1:ZBox 2:Zoom In 3:Zoom Out 4:ZDecimal 5:ZSquare 6:ZStandard 7↓ZTrig```

Press WINDOW to see the view set by ZDecimal.

Xmin and Xmax are the settings of the left and right edges of the viewing screen, and Ymin and Ymax are the settings for the lower and upper edges of the viewing screen.

Xscl and Yscl set the spacing between the tick marks on the x- and y-axes.

The view you see is $-4.7 \le x \le 4.7$, $-3.1 \le y \le 3.1$.

```
WINDOW FORMAT
 Xmin= -4.7
 Xmax=4.7
 Xscl=1
 Ymin= -3.1
 Ymax=3.1
 Yscl=1
```

Follow the procedures shown in either 1.1.3 or 1.1.4 to draw a graph with your calculator. Whenever you draw a graph, you have the option of manually changing the view or having the calculator automatically find a view of the graph.

1.1.3 MANUALLY CHANGING THE VIEW OF A GRAPH   If you do not have a good view of the graph or if you do not see the graph, change the view with one of the ZOOM options or manually set the WINDOW. (We later discuss the ZOOM options.)

TI-82 Press WINDOW ▼, and set Xmin to 0, Xmax to 10, leave Xscl at 1, set Ymin to 900, Ymax to 2000, and Yscl to 50.

```
WINDOW FORMAT
 Xmin=0
 Xmax=10
 Xscl=1
 Ymin=900
 Ymax=2000
 Yscl=50
```

TI-83 Press WINDOW and set Xmin to 0, Xmax to 10, leave Xscl at 1, set Ymin to 900, Ymax to 2000, and Yscl to 50.

For all applications in this *Guide*, have Xres set to 1.

```
WINDOW
 Xmin=0
 Xmax=10
 Xscl=1
 Ymin=900
 Ymax=2000
 Yscl=50
 Xres=1
```

Both Press GRAPH to draw the graph of $Y1 = 1000(1 + 0.05)^\wedge X$ using the new view.

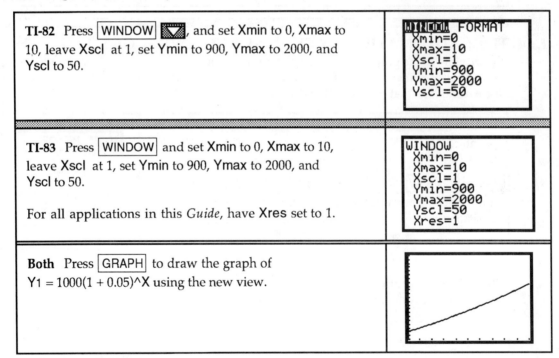

1.1.4 TI-82: AUTOMATICALLY CHANGING THE VIEW OF A GRAPH

If your view of the graph is not good or if you do not see the graph, change the view using the TI-82 program AUTOSCL. If you do not have this program, either enter it now using the code given in the TI-82 Appendix or have someone transfer it to you from another calculator.

Caution: When using this TI-82 program, the function you are graphing must be entered in the Y1 location of the Y= graphing list. If other functions are entered in the graphing list, this program will *not* find an appropriate view for all the functions.

$Y1 = 1000(1 + 0.05)^x$

(Delete other functions that may be entered in the other locations in the Y= list.)

Press PRGM followed by the number corresponding to the location of program AUTOSCL. Press ENTER. *Note:* Your program list may or may not look exactly like the one that is shown to the right.	EXEC EDIT NEW 1:AUTOSCL 2:DIFF 3:EULER 4:LOGISTIC 5:LSLINE 6:NUMINTGL 7↓SECTAN
You are asked to enter **Xmin**, the value at which the left side of the view is set. Type 0 and press ENTER. Next, you are asked to enter **Xmax**, the value at which the right side of the view is set. Type 10 and press ENTER.	HAVE F(X) IN Y1 Xmin? 0 Xmax? 10
Program **AUTOSCL** now draws the graph. *Note:* In some cases, the graph drawn by the program will not be the best view. If this happens, follow the instructions in part 8 to manually set **Ymin** and **Ymax** (and/or the scale values) to obtain a better view.	

1.1.4 TI-83: AUTOMATICALLY CHANGING THE VIEW OF THE GRAPH

If your view of the graph is not good or if you do not see the graph, change the view using the built-in autoscaling feature of the TI-83. This option will automatically find a view to see all the functions that you have turned on in the graphing list.

Be sure the function you are graphing, $y = 1000(1 + 0.05)^x$, is entered in the **Y1** location of the **Y=** list. (Delete all other functions that may be entered in other locations.) Press WINDOW, set **Xmin** to 0 and **Xmax** to 10. (It does not matter what values are set in the **Ymin** and **Ymax** positions.) Press ZOOM ▲ (0: ZoomFit) ENTER.	ZOOM MEMORY 4↑ZDecimal 5:ZSquare 6:ZStandard 7:ZTrig 8:ZInteger 9:ZoomStat 0:ZoomFit
The TI-83 automatically sets the vertical view and (based on the **Xmin** and **Xmax** you set) draws a graph of the function.	

1.1.5 TRACING
You can display the coordinates of certain points on the graph by tracing. The x-values shown when you trace depend on the horizontal view that you choose, and the y-values are calculated by substituting the x-values into the equation that is being graphed.

TI-82 Press TRACE, use ▶ to move the trace cursor to the right, and use ◀ to move the trace cursor to the left. The number 1 in the upper right hand corner of the screen tells you that you are tracing on the equation in **Y**1.	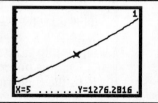 X=5 ......Y=1276.2816

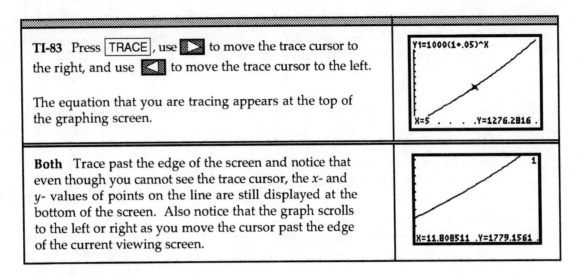

TI-83 Press ⌈TRACE⌋, use ▶ to move the trace cursor to the right, and use ◀ to move the trace cursor to the left.

The equation that you are tracing appears at the top of the graphing screen.

Y1=1000(1+.05)^X

X=5 . . . .Y=1276.2816 .

Both Trace past the edge of the screen and notice that even though you cannot see the trace cursor, the *x*- and *y*- values of points on the line are still displayed at the bottom of the screen. Also notice that the graph scrolls to the left or right as you move the cursor past the edge of the current viewing screen.

X=11.808511 .Y=1779.1561 .

1.1.6 ESTIMATING OUTPUTS   You can estimate outputs from the graph using TRACE. It is important to realize that such outputs are *never* exact values unless the displayed *x*-value is *identically* the same as the value of the input variable.

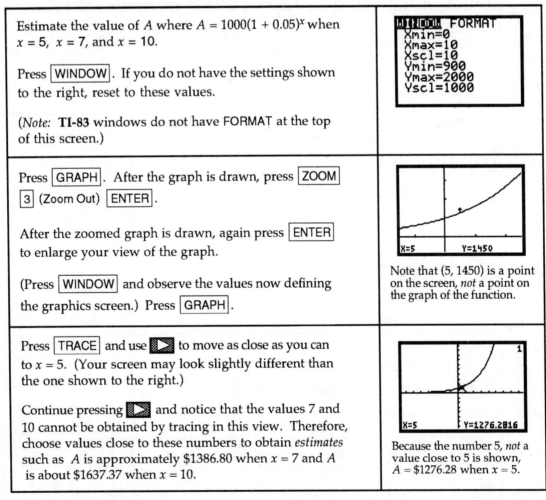

Estimate the value of *A* where $A = 1000(1 + 0.05)^x$ when $x = 5$, $x = 7$, and $x = 10$.

Press ⌈WINDOW⌋. If you do not have the settings shown to the right, reset to these values.

(*Note:* **TI-83** windows do not have FORMAT at the top of this screen.)

WINDOW FORMAT
Xmin=0
Xmax=10
Xscl=10
Ymin=900
Ymax=2000
Yscl=1000

Press ⌈GRAPH⌋. After the graph is drawn, press ⌈ZOOM⌋ ⌈3⌋ (Zoom Out) ⌈ENTER⌋.

After the zoomed graph is drawn, again press ⌈ENTER⌋ to enlarge your view of the graph.

(Press ⌈WINDOW⌋ and observe the values now defining the graphics screen.) Press ⌈GRAPH⌋.

X=5   Y=1450

Note that (5, 1450) is a point on the screen, *not* a point on the graph of the function.

Press ⌈TRACE⌋ and use ▶ to move as close as you can to $x = 5$. (Your screen may look slightly different than the one shown to the right.)

Continue pressing ▶ and notice that the values 7 and 10 cannot be obtained by tracing in this view. Therefore, choose values close to these numbers to obtain *estimates* such as *A* is approximately $1386.80 when $x = 7$ and *A* is about $1637.37 when $x = 10$.

X=5   Y=1276.2816

Because the number 5, *not* a value close to 5 is shown, $A = 1276.28 when $x = 5$.

- If you had used the original WINDOW with Xmax = 10 and traced, you should obtain the *exact* value $A = 1500$ when $x = 10$ because 10, not a value close to 10, is shown when tracing.

- If you want "nice, friendly" values displayed for x when tracing, set Xmin and Xmax so that Xmax − Xmin is a multiple of 9.4, the width of the ZDecimal viewing screen. For instance, if you set Xmin = 0 and Xmax = 18.8 in the example above, the *exact* values when $x = 5$, $x = 7$, and $x = 10$ are displayed when you trace since 18.8 = 2(9.4). Try it!

1.1.7 EVALUATING OUTPUTS

The values obtained by this evaluation process are *actual* output values of the equation, not estimated values such as those generally obtained by tracing. Begin by entering the equation whose output you want to evaluate in the Y= list.

Using x as the input variable, enter the function in the Y= list. Return to the home screen by pressing [2nd] [MODE] (QUIT) as many times as necessary.	Y1= $1000(1 + 0.05)^x$ (You could use any of the 10 function locations.)
TI-82 Go to the Y-VARS menu by pressing [2nd] [VARS] (Y-VARS).	Y-VARS 1 Function… 2:Parametric… 3:Polar… 4:Sequence… 5:On/Off…
TI 83 Go to the Y-VARS menu by pressing [VARS] [▶] (Y-VARS).	VARS Y-VARS 1 Function… 2:Parametric… 3:Polar… 4:On/Off…
Both Choose 1: Function by pressing [1] or [ENTER], and choose 1: Y1 by pressing [1] or [ENTER]. (To choose another Y= location, simply press the number corresponding to that function.)	FUNCTION 1 Y1 2:Y2 3:Y3 4:Y4 5:Y5 6:Y6 7↓Y7
Y1 shows on your screen. Type the x-value at which you want to evaluate the equation, and press [ENTER]. Now, evaluate Y1 at $x = 5$. (*Note:* You can, but do not have to, type in the *closing* parentheses on the right.)	Y1(5) 1276.281563

Evaluate Y1 at $x = 7$ by recalling the previous entry with [2nd] [ENTER] (ENTRY), edit the 5 to 7 by pressing [◀] [◀] and typing over the 5, and press [ENTER].	Y1(5)
	         1276.281563
	Y1(7)
	         1407.100423

Evaluate Y1 at $x = 10$ by recalling the previous entry with [2nd] [ENTER] (ENTRY), edit the 7 to 10 by pressing [◀] [◀] and typing over the 7, and press [ENTER].	Y1(5)
	         1276.281563
	Y1(7)
	         1407.100423
	Y1(10
	         1628.894627

📖 1.2 Functions and Graphs

When you are asked to *estimate* or *approximate* an output or an input value, you can use your calculator in the following ways:

- tracing a graph   (Sections 1.1.5, 1.1.6)
- close values obtained from a table of function values   (End of Section 1.2.2)

When you are asked to *find* or *determine* an output or an input value, you should use your calculator in the following ways:

- evaluating an output on the home screen   (Section 1.1.7)
- find a value using the **AUTO** or **ASK** features of the table   (Section 1.2.1)
- determine an input using the solver   (Section 1.2.2)

1.2.1 DETERMINING OUTPUTS

Function outputs can be determined by evaluating on the home screen, as discussed in 1.1.7. You can also evaluate functions using the calculator's **TABLE**. When you use the table, you can ask for specific output values corresponding to the inputs you enter or generate a list of input values that begin with TblMin and differ by ΔTbl and their corresponding outputs.

Let's use the **TABLE** to determine the output of the function $v(t) = 3.622(1.093)^t$ when $t = 85$. Even though you could use any of the function locations, we choose to use Y1. Press $\boxed{Y=}$, clear the function locations, and enter 3.622(1.093)^X in location Y1 of the Y= list.

After entering the function $v(t)$ in Y1, choose the **TABLE SETUP** menu.	Press $\boxed{2nd}$ $\boxed{WINDOW}$ (TblSet).
To generate a list of values beginning with an input of 80 with the table values differing by 1, enter 80 in the TblMin location, 1 in the ΔTbl location, and choose AUTO in the Indpnt: and Depend: locations. Remember that you "choose" a particular setting by positioning the blinking cursor over that setting and pressing $\boxed{ENTER}$.	TABLE SETUP TblMin=80 ΔTbl=1 Indpnt: **Auto** Ask Depend: **Auto** Ask
Press $\boxed{2nd}$ $\boxed{GRAPH}$ (TABLE), and observe the list of input and output values. Notice that you can scroll through the table with ▼, ▶, ▲, and/or ◀. The table values may be rounded in the table display. You can see more of the output by moving to the value and looking at the bottom of the screen.	X | Y1 80 | 4452.1 81 | 4866.1 82 | 5318.7 83 | 5813.3 84 | 6353.9 85 | **6944.8** 86 | 7590.7 Y1=6944.83947371
Return to the **TABLE SETUP** menu with $\boxed{2nd}$ $\boxed{WINDOW}$ (TblSet). To compute specific outputs rather than a list of values, choose ASK in the Indpnt: location. (When using ASK, the settings for TblMin and ΔTbl do not matter.)	TABLE SETUP TblMin=80 ΔTbl=1 Indpnt: Auto **Ask** Depend: **Auto** Ask

Press 2nd GRAPH (TABLE), type in the *x*-value(s) at which the function is to be evaluated, and press ENTER.

You can scroll through the table with ▼, ▶, ▲, and/or ◀. Unwanted input entries can be cleared with DEL.

Using any of these methods, we see that $v(85) \approx \$6945$.

X	Y₁	
85	6944.8	

Y₁=6944.83947371

1.2.2 SOLVING FOR INPUT VALUES

Your calculator solves for the input values of any equation that you have put in the form *"expression = 0"*. The expression can, but does not have to, use X as the input variable. However, you must specify the variable you are using.

TI-82 The form of the expression you type in the solver is *Solve(expression, variable, initial guess)*.	You will use the **solve** command throughout this course.
Press 2nd MODE (QUIT) to return to the home screen. Access the **MATH** menu with MATH. Use ▼ to locate 0: solve(.	**MATH** NUM HYP PRB 4↑³√ 5:×√ 6:fMin(7:fMax(8:nDeriv(9:fnInt(0▮solve(
TI-83 The TI-83 gives you two methods of solving for input variables. **METHOD 1:** The expression you type in your calculator is of the form *Solve(expression, variable, initial guess)*.	This method is very similar to that for the TI-82. The main difference is in the location of the **solve** command.

Return to the home screen with 2nd MODE (QUIT).
 	
Access the **CATALOG** menu with 2nd 0 (CATALOG). Press LN to access the list of commands beginning with S (because LN is the key with S written above it). Use ▼ to locate **solve(**.
	CATALOG ▮
	  Simul
	  sin(
	  sin⁻¹(
	  sinh(
	  sinh⁻¹(
	  SinReg
	▶solve(

Both Press ENTER to copy the instruction to the home screen. Suppose we want to solve $v(t) = 3.622(1.093)^t$ for *t* when $v = \$15,000$.	solve(
Since the equation you enter is *"expression = 0"*, subtract 15,000 from both sides of the equation to obtain $$0 = 15,000 - 15,000 = 3.622(1.093)^t - 15,000$$ Tell the calculator the variable with ALPHA 4 (T), provide a guess, and press ENTER.	solve(3.622(1.09 3)^T-15000,T,50) 93.65944187

Note that your *guess* can be obtained by drawing a graph of $3.622(1.093)^t$ and tracing until you have an estimate of where the output is 15,000. (Remember that to graph any function, you must rewrite it so that the input variable is x.)	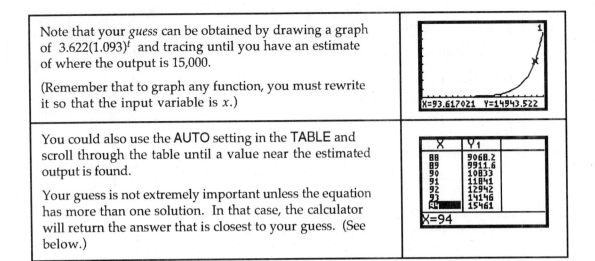
You could also use the AUTO setting in the TABLE and scroll through the table until a value near the estimated output is found. Your guess is not extremely important unless the equation has more than one solution. In that case, the calculator will return the answer that is closest to your guess. (See below.)	

If there is more than one solution to an equation, you need to give the solver an approximate location for each answer. Suppose you are given $q(x) = 8x^2 + 54.65x - 163$ and asked to find what input(s) correspond to an output of $q(x) = 154$. (The procedure outlined below also applies to finding where two functions are equal.)

Enter $8x^2 + 54.65x - 163$ in one location, say Y1, and 154 in another location, say Y2, in the Y= list. (Remember that if the input variable in the equation is not x, you must rewrite the equation in terms of x to graph using the Y= list.)	
To better obtain a guess as to where Y1 equals (intersects) Y2, graph the equations. If you are not told where you want to view the graph, begin by pressing ZOOM 4 (ZDecimal) or ZOOM 6 (ZStandard). You want to see a "good" graph, that is, one that shows all the important features. In this case, the important features are where Y1 and Y2 intersect. The top graph was obtained with ZDecimal and the bottom graph with ZStandard. Neither graph on the right is a good graph for viewing the intersections.	
To improve the view, press WINDOW and change the settings to Xmin = ⁻15, Xmax = 8, Ymin = ⁻300, and Ymax = 400 . Draw the graph with GRAPH . (There are many other windows that work just as well as the one shown to the right. Also, instead of setting a window manually, you could use program AUTOSCL on the TI-82 or use ZoomFit on the TI-83.)	

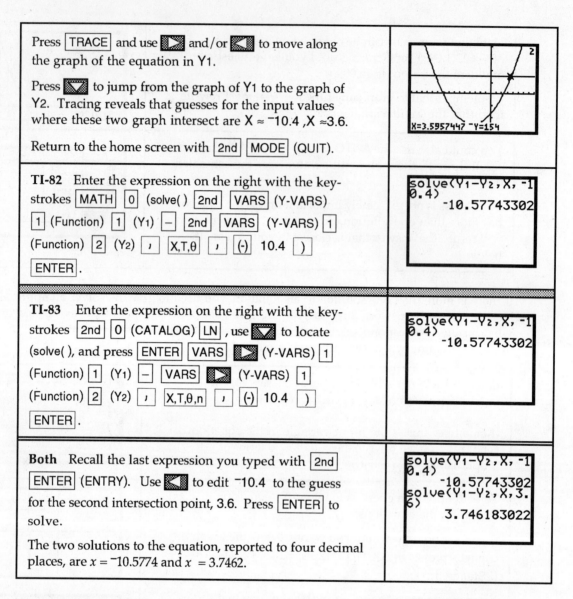

Press TRACE and use ▶ and/or ◀ to move along the graph of the equation in Y1.

Press ▼ to jump from the graph of Y1 to the graph of Y2. Tracing reveals that guesses for the input values where these two graph intersect are X ≈ ⁻10.4 , X ≈ 3.6.

Return to the home screen with 2nd MODE (QUIT).

TI-82   Enter the expression on the right with the keystrokes MATH 0 (solve( ) 2nd VARS (Y-VARS) 1 (Function) 1 (Y1) – 2nd VARS (Y-VARS) 1 (Function) 2 (Y2) , X,T,θ , (-) 10.4) ENTER.

TI-83   Enter the expression on the right with the keystrokes 2nd 0 (CATALOG) LN , use ▼ to locate (solve( ), and press ENTER VARS ▶ (Y-VARS) 1 (Function) 1 (Y1) – VARS ▶ (Y-VARS) 1 (Function) 2 (Y2) , X,T,θ,n , (-) 10.4) ENTER.

Both   Recall the last expression you typed with 2nd ENTER (ENTRY). Use ◀ to edit ⁻10.4 to the guess for the second intersection point, 3.6. Press ENTER to solve.

The two solutions to the equation, reported to four decimal places, are $x =$ ⁻10.5774 and $x = 3.7462$.

The TI-83 offers a second method of solving that you may find more convenient than the method previously discussed.

TI-83   This second method of solving can be used instead of the first method anytime you need to solve for an input variable or find where two expressions are equal.

METHOD 2:   The equation containing the unknown quantity must have all terms on the left-hand side and a 0 on the right-hand side or vice-versa.)

This method uses the equation solver that is built into the TI-83.

You must, like in the first method, enter an expression that equals 0.

Return to the home screen with 2nd MODE (QUIT).

Access the **solver** by pressing MATH 0 . If there are no equations stored in the **solver**, you will see the screen displayed on the right -- or

```
EQUATION SOLVER
eqn:0=
```

-- if the solver has been used previously, you will see a screen something like the one shown on the right. If this is the case, press ▲ CLEAR and you should then have the screen shown in the previous box.	``` Y₁-23=0 X=-57.9787234 bound=(-1E99,1... ```
Let us return to the example where we are solving $v(t) = 3.622(1.093)^t$ for t when $v = \$15{,}000$. Since the equation you enter is "eqn = 0", subtract 15,000 from both sides of the equation $15{,}000 = 3.622(1.093)^t$, and enter the expression shown on the right.	``` EQUATION SOLVER eqn:0=3.622(1.09 3)^T-15000 ```
Press ENTER or ▼. (Note: To return to the equation for editing purposes, press ▲ until the previous screen reappears.) With the cursor covering the 5 in the T location, enter a guess, say 10. (You could use the default guess that automatically appears.)	``` 3.622(1.093)^...=0 T=5 bound=(-1E99,1... ```
Press ENTER or ▼ and the cursor moves to the bound location. These are the values between which the TI-83 searches for a solution. You do not have to change these values unless you get an error message or you want a quicker search for the solution. *Therefore, you can often skip this step.*	``` 3.622(1.093)^...=0 T=10 bound=■-1E99,1... ```
Move the cursor to the location of the variable for which you are solving. (In this example, T.) Press ALPHA ENTER (SOLVE). (Notice the dot that appears next to the T and by the last line on this screen. This means a solution has been found.) The bottom line on the screen, "left-rt = 0", indicates the value found for T is an exact solution since both sides of the equation evaluate to the same quantity.	``` 3.622(1.093)^...=0 •T=93.659441868... bound=(-1E99,1... •left-rt=0 ```

TI-83 Note: Remember the following when using the equation solver:

- You can use functions stored in the Y= list in the SOLVER, but you must rewrite the function so that x is the input variable. At the eqn: 0= prompt, enter the location of the function using the VARS (Y-VARS) menu. For instance, if Y1 = 3.622(1.093)^X, enter Y1 −15000 after the eqn: 0= in the equation solver.

- If a solution continues beyond the edge of the screen, you see "..." to the right of the value. Be certain that you use ▶ to scroll to the end of the number. The value may be given in scientific notation, and the portion that you cannot see determines the location of the decimal point. (See *Basic Operation, #3,* of this *Guide*.)

1.2.3 GRAPHICALLY FINDING INTERCEPTS   Finding where a function graph crosses the vertical and horizontal axis can be done graphically as well as by the methods indicated in 1.2.2 of this *Guide*. Remember the process by which we find intercepts:

- To find the *y*-intercept of a function $y = f(x)$, set $x=0$ and solve the resulting equation.

- To find the *x*-intercept of a function $y = f(x)$, set $y=0$ and solve the resulting equation.

Also remember that an *x*-intercept of a function $y = f(x)$ has the same value as the root or solution of the equation $f(x) = 0$.

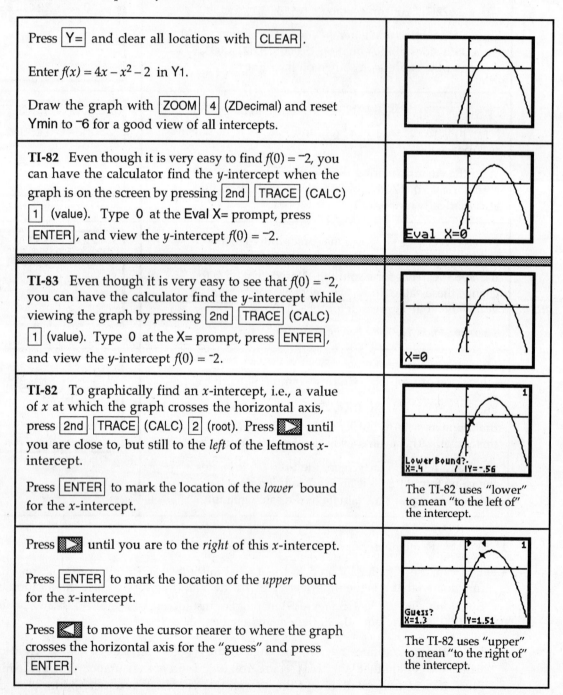

Press Y= and clear all locations with CLEAR .

Enter $f(x) = 4x - x^2 - 2$ in Y1.

Draw the graph with ZOOM 4 (ZDecimal) and reset Ymin to ‾6 for a good view of all intercepts.

TI-82 Even though it is very easy to find $f(0) = ‾2$, you can have the calculator find the *y*-intercept when the graph is on the screen by pressing 2nd TRACE (CALC) 1 (value). Type 0 at the Eval X= prompt, press ENTER , and view the *y*-intercept $f(0) = ‾2$.

Eval X=0

TI-83 Even though it is very easy to see that $f(0) = ‾2$, you can have the calculator find the *y*-intercept while viewing the graph by pressing 2nd TRACE (CALC) 1 (value). Type 0 at the X= prompt, press ENTER , and view the *y*-intercept $f(0) = ‾2$.

X=0

TI-82 To graphically find an *x*-intercept, i.e., a value of *x* at which the graph crosses the horizontal axis, press 2nd TRACE (CALC) 2 (root). Press ▶ until you are close to, but still to the *left* of the leftmost *x*-intercept.

Press ENTER to mark the location of the *lower* bound for the *x*-intercept.

Lower Bound?.
X=.4   / IY=‾.56

The TI-82 uses "lower" to mean "to the left of" the intercept.

Press ▶ until you are to the *right* of this *x*-intercept.

Press ENTER to mark the location of the *upper* bound for the *x*-intercept.

Press ◀ to move the cursor nearer to where the graph crosses the horizontal axis for the "guess" and press ENTER .

Guess?
X=1.3   Y=1.51

The TI-82 uses "upper" to mean "to the right of" the intercept.

The value of the leftmost *x*-intercept is displayed as
X = 0.58578644.

Repeat the above procedure to find the other *x*-intercept.
Confirm that it is X = 3.4142136.

Root
X=.58578644 Y=0

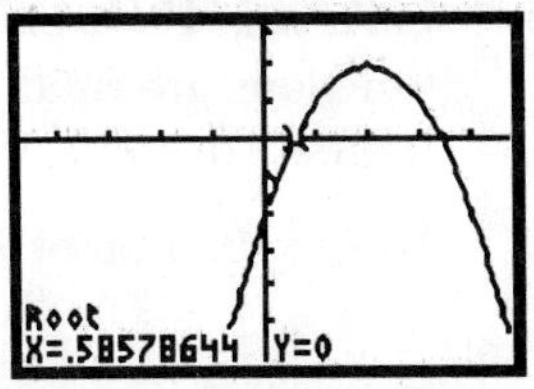

TI-83   To graphically find an *x*-intercept, i.e., a value
of *x* at which the graph crosses the horizontal axis,
press 2nd TRACE (CALC) 2 (zero).

Press ▶ until you are close to, but still to the *left* of
the leftmost *x*-intercept. Press ENTER to mark the
location of the *left* bound for the *x*-intercept.

Y1=4X-X2-2

Left Bound?
X=.3   Y=-.89

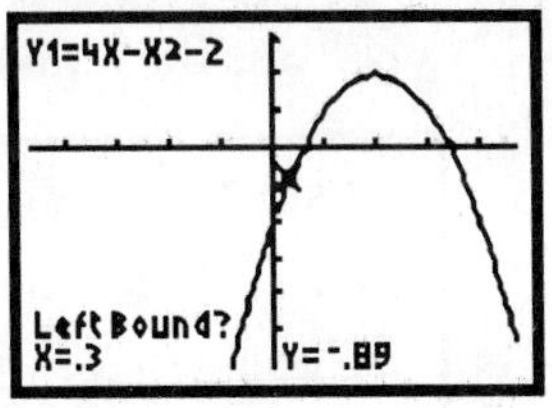

Press ▶ until you are to the *right* of this *x*-intercept.
Press ENTER to mark the location of the *right* bound
for the *x*-intercept.

For your "guess", press ◀ to move the cursor nearer to
where the graph crosses the horizontal axis and press
ENTER.

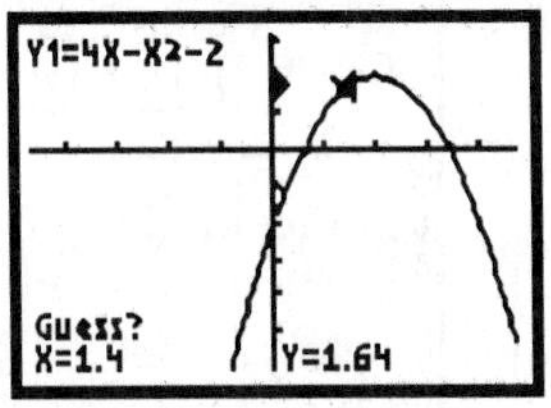

Y1=4X-X2-2

Guess?
X=1.4   Y=1.64

The value of the leftmost *x*-intercept is displayed as
X = 0.58578644.

Repeat the above procedure to find the other *x*-intercept.
Confirm that it is X = 3.4142136.

Zero
X=.58578644 Y=0

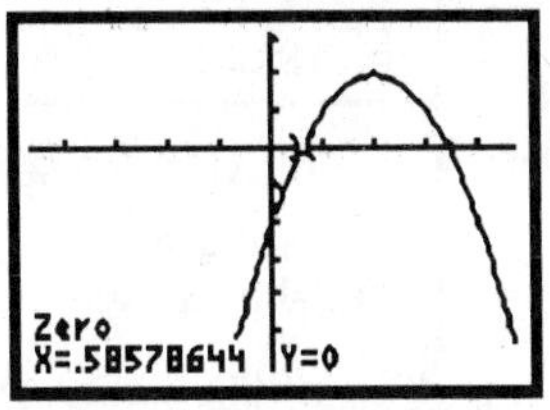

1.3 Constructed Functions

Your calculator can find output values of and graph combinations of functions in the same
way that you do these things for a single function. The only additional information you need
is how to enter constructed functions in the graphing list. Suppose that a function *f(x)* has
been entered in Y1 and a function *g(x)* has been entered in Y2.

- Enter Y1 + Y2 in Y3 to obtain the sum function $(f+g)(x) = f(x) + g(x)$.

- Enter Y1 − Y2 in Y4 to obtain the sum function $(f–g)(x) = f(x) –g(x)$.

- Enter Y1*Y2 in Y5 to obtain the product function $(f·g)(x) = f(x) * g(x)$.

- Enter Y1/Y2 in Y6 to obtain the quotient function $(f÷g)(x) = \dfrac{f(x)}{g(x)}$.

- Enter Y1(Y2) in Y7 to obtain the composite function $(f∘g)(x) = f(g(x))$.

Your calculator will evaluate and graph these constructed functions. Although it will not
give you an algebraic formula for a constructed function, you can check your algebra by
evaluating the calculator-constructed function and your constructed function at several
different points.

1.3.1 GRAPHING PIECEWISE CONTINUOUS FUNCTIONS

Piecewise continuous functions are used throughout the text. It is often helpful to use your calculator to graph and evaluate outputs of piecewise continuous functions. Consider the following example.

The population of West Virginia from 1985 through 1993 can be modeled by

$$P(t) = \begin{cases} -23.514t + 3903.667 \text{ thousand people when } 85 \le t < 90 \\ 9.1t + 972.6 \text{ thousand people when } 90 \le t \le 93 \end{cases}$$

where t is the number of years since 1900.

TI-82 Enter the function $P(t)$, using x as the input variable, in the Y1 location of the Y= list using the keystrokes [(] [(-)] 23.514 [X,T,θ] [+] 3903.667 [)] [(] [X,T,θ] [2nd] [MATH] (TEST) [5] (<) 90 [)] [+] [(] 9.1 [X,T,θ] [+] 972.6 [)] [(] [X,T,θ] [2nd] [MATH] (TEST) [4] (≥) 90 [)]	Y1◻(-23.514X+390 3.667)(X<90)+(9. 1X+972.6)(X≥90) Y2= Y3= Y4= Y5= Y6=
Notice that the function is defined only when the input is between 85 and 93. You could find $P(85)$ and $P(93)$ to help you set the vertical view. However, we choose to use program **AUTOSCL**. (See Section 1.1.4 of this *Guide*.) When prompted, enter Xmin = 85 and Xmax = 93. The graph to the right is drawn by the program.	
If you wish to see the "break" in the function where the two pieces join, the width of the screen must be a multiple of 9.4 and include 90. Since $90 - 0.5(9.4) = 85.3$ and $90 + 0.5(9.4) = 94.7$, change Xmin and Xmax to these values and press [GRAPH].	 X=90.......Y=1791.6 ...
Because the two pieces are close together at $x = 90$, you may need to take a closer look to see the break. However, because the calculator draws graphs by connecting function outputs wherever the function is defined, it will connect the two pieces unless you tell it not to do so by pressing [MODE], use [▼] and [▶] to choose Dot , and press [ENTER].	Normal Sci Eng Float 0123456789 Radian Degree Func Par Pol Seq Connected Dot Sequential Simul FullScreen Split
Now, take a closer look with [GRAPH] and [ZOOM] [2] (Zoom In). To keep the point where the functions break in view, use [▼] to move the small cursor that appears in the middle of the screen down to where the two functions join before pressing [ENTER] to actually zoom in.	 X=90 .Y=1791.7685

You can find function values by evaluating outputs on the home screen or using the table.
 	
Do not forget to change the calculator's MODE setting back to **Connected** when you finish graphing the piecewise function.
	Y1(87)
	      1857.949
	Y1(92)
	        1809.8
	Y1(85)
	      1904.977
	Y1(90)■

TI-83 The same procedure as given in the above steps for the TI-82 will also work for the TI-83. However, the TI-83 has some additional features that will make graphing of piecewise functions less complicated.

(See Section 6.1.2 of this *Guide* for an example showing how to fit a piecewise model to data.)

Enter the function $P(t)$, using x as the input variable, in the Y1 location of the Y= list using the keystrokes [( ]

[(−)] 23.514 [X,T,θ,n] [+] 3903.667 [ )] [( ] 85 [2nd]

[MATH] (TEST) [6] (≤) [X,T,θ,n] [ )] [( ] [X,T,θ,n] [2nd]

[MATH] (TEST) [5] (<) 90 [ )] [+] [( ] 9.1 [X,T,θ,n] [+]

972.6 [ )] [( ] [X,T,θ,n] [2nd] [MATH] (TEST) [4] (≥) 90

[ )] [( ] [X,T,θ,n] [2nd] [MATH] (TEST) [6] (≤) 93 [ )]

Notice that these keystrokes specify the complete input of the function P.

```
Plot1 Plot2 Plot3
\Y1◙(-23.514X+39
03.667)(85≤X)(X<
90)+(9.1X+972.6)
(X≥90)(X≤93)
\Y2=
\Y3=
\Y4=
```

Each piece of the function and its corresponding input must be enclosed in parentheses.

This piecewise function "breaks" at $x = 90$. However, your calculator draws graphs by connecting function outputs wherever the function is defined, so it will connect the two pieces unless you tell it not to do. Whenever you draw graphs of piecewise functions, it is easier to set your calculator to Dot mode in the following manner.

Have the cursor on the first line of the function and press [◀] until you highlight the slanted line* to the *left* of Y1. Press [ENTER] six times.

```
Plot1 Plot2 Plot3
∵Y1◙(-23.514X+39
03.667)(85≤X)(X<
90)+(9.1X+972.6)
(X≥90)(X≤93)
\Y2=
\Y3=
\Y4=
```

The dotted line to the left of Y1 indicates the graph will draw without joining the outputs of the function.

Notice that the function is defined only when the input is between 85 and 93. You could find $P(85)$ and $P(93)$ to help you set the vertical view. However, we choose to use ZoomFit. (See Section 1.1.4 of this *Guide*.)

Press [WINDOW], set Xmin to 85, and set Xmax to 93. Press [ZOOM] [▲] (0: ZoomFit) [ENTER].

(The breaks you see in the left portion of the graph are because you have told the calculator not to connect outputs.)

Even though you can see the "break" in the function where the two pieces join, you cannot trace to that point. To do this, the width of the screen must be a multiple of 9.4 and must include 90.

Since $90 - 0.5(9.4) = 85.3$ and $90 + 0.5(9.4) = 94.7$, change Xmin and Xmax to these values. Also change Ymin to 1750. Press [GRAPH].

```
Y1=(-23.514X+3903.667)(8_

X=90       .Y=1791.6  .
```

---

* The different "graph styles" you can draw from this location are described in more detail on pages 3-9 through 3-10 in your TI-83 Owner's *Guidebook*.

You can find function values by evaluating outputs on the home screen or using the table.

```
Y₁(87)
              1857.949
Y₁(92)
              1809.8
Y₁(85)
              1904.977
Y₁(90)■
```

1.4 Linear Functions and Models

Actual real-world data is used throughout *Calculus Concepts*. It is necessary that you use your calculator to find a curve that models the data. Be very careful when you enter the data in your calculator because your model and all of your results depend on the values that you enter!

1.4.1 ENTERING DATA   Press [STAT] [1] (EDIT) to access the six lists that hold data. You only see the first three lists, L1, L2, and L3, but you can access the other three, L4, L5, and L6, with ▶. (In this text, we usually use list L1 for the input data and list L2 for the output data.)   If there are any data values in your lists, see 1.4.3 of this *Guide* and first delete any "old" data. (**TI-83** Note: If you do not see L1, L2, and so forth, return to the statistical setup instructions at the beginning of this *Guide*.)

Enter the following data:

Year	1992	1993	1994	1995	1996	1997
Tax	2541	3081	3615	4157	4703	5242

Position the cursor in the first location in list L1. Enter the *x*-data into list L1 by typing the entries from top to bottom in the L1 column, pressing [ENTER] after each entry.

After typing the L1(6) value, 1997, use ▶ to go to the top of list L2. Enter the *y*-data into list L2 by typing the entries from top to bottom in the L2 column, pressing [ENTER] after each entry.

1.4.2 EDITING DATA   If you incorrectly type a data value, use the cursor keys to darken the value you wish to correct and type the correct value. Press [ENTER].

- To *insert* a data value, put the cursor over the value that will be directly below the one you will insert, and press [2nd] [DEL] (INS). The values in the list below the insertion point move down one location and a 0 is filled in at the insertion point. Type the data value to be inserted over the 0 and press [ENTER]. The 0 is replaced with the new value.

- To *delete* a single data value, move the cursor over the value you wish to delete, and press [DEL]. The values in the list below the deleted value move up one location.

1.4.3 DELETING OLD DATA   Whenever you enter new data in your calculator, you should first delete any previously-entered data. There are several ways to do this, and the most convenient method is illustrated below.

Access the data lists with $\boxed{\text{STAT}}$ $\boxed{1}$ (EDIT). (You probably have different values in your lists if you are deleting "old" data.) Use $\boxed{\blacktriangle}$ to move the cursor over the name L1.	L1 L2 L3 0 37 ------ 1 35 3 29 6 20 8 14 ------ ------ L1=(0,1,3,6,8)

Press $\boxed{\text{CLEAR}}$ $\boxed{\text{ENTER}}$.	L1 L2 L3
	 ■■■■ 37 ------
	       35
	       29
	       20
	       14
	       ------ 	
	 L1(1)=

Use $\boxed{\blacktriangleright}$ $\boxed{\blacktriangle}$ to move the cursor over the name L2. Press $\boxed{\text{CLEAR}}$ $\boxed{\text{ENTER}}$. Repeat this procedure to clear data from any of the other lists you want to use.	L1 L2 L3 ------ ------ ------ L2=

1.4.4 ALIGNING DATA   Let's now return to the data you entered in Section 1.4.1 of this *Guide*. Suppose you want L1 to contain the number of years since a certain year (here, 1992) instead of actual years. That is, you want to *align* the x-data. In this example, you are to shift all the data values 1992 units to the left of where they currently are located.

Position the cursor over the L1 at the top of the first column. Replace the L1 values with $L1 - 1992$ values by pressing $\boxed{\text{2nd}}$ $\boxed{1}$ (L1) $\boxed{-}$ 1992 $\boxed{\text{ENTER}}$. Instead of an actual year, the input now represents the number of years after 1992.	L1 L2 L3 0 2541 ------ 1 3081 2 3615 3 4157 4 4703 5 5242 ------ ------ L1=L1−1992

1.4.5 PLOTTING DATA   Any functions you have in the Y= list will graph when you plot data. Therefore, you should clear or turn them off before drawing a scatter plot.

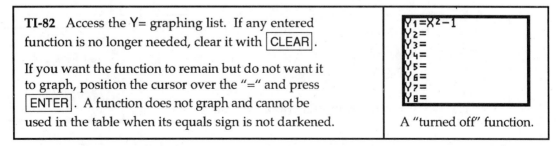

TI-82 Access the Y= graphing list. If any entered function is no longer needed, clear it with $\boxed{\text{CLEAR}}$. If you want the function to remain but do not want it to graph, position the cursor over the "=" and press $\boxed{\text{ENTER}}$. A function does not graph and cannot be used in the table when its equals sign is not darkened.	Y1=X²−1 Y2= Y3= Y4= Y5= Y6= Y7= Y8= A "turned off" function.

Press 2nd Y= (STAT PLOT) to display the STAT PLOTS screen.

Note: When drawing a graph from the Y= list, you may get an error message or see a scatter plot of "old" data as well as the function graph. If so, turn off the STAT PLOTS with 4 (PlotsOff) ENTER.

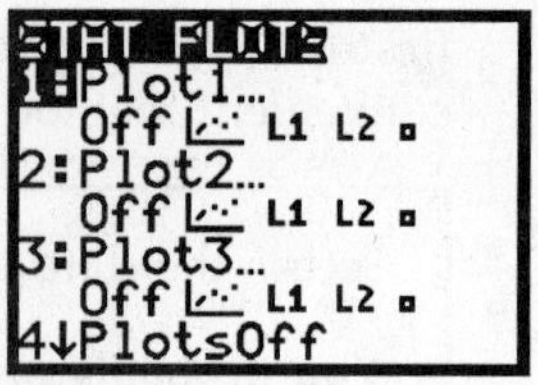

On the STAT PLOTS screen, press ENTER to display the Plot1 screen, press ENTER to turn Plot1 "On", and select the options shown on the right.

(You can choose any of the three marks at the bottom of the screen.)

Press ZOOM 9 (ZoomStat) to have the calculator set an autoscaled view of the data and draw the scatter plot.

(ZoomStat does not reset the x and y-axis tick marks. You should do this manually with WINDOW if you want different spacing between the marks.)

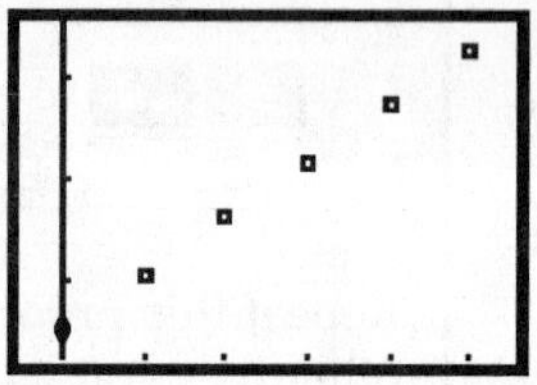

TI-83 Access the Y= graphing list. If any entered function is no longer needed, clear it with CLEAR.

If you want the function to remain but do not want it to graph, position the cursor over the "=" and press ENTER. A function does not graph and cannot be used in the table when its equals sign is not darkened.

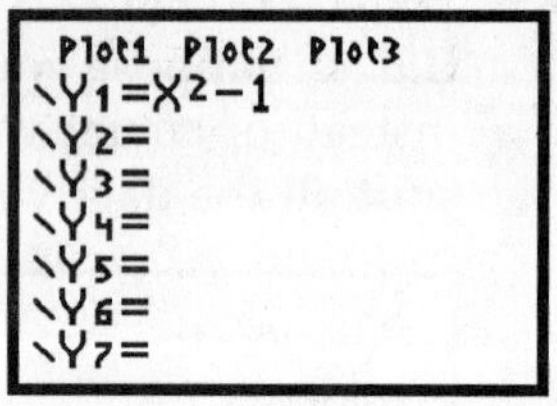

A "turned off" function.

Press 2nd Y= (STAT PLOT) to display the STAT PLOTS screen. (Your screen may not look exactly like this one.)

Note: When drawing a graph from the Y= list, you may get an error message or see a scatter plot of "old" data as well as the function graph. If so, turn off the STAT PLOTS with 4 (PlotsOff) ENTER.

On the STAT PLOTS screen, press ENTER to display the Plot1 screen, press ENTER to turn Plot1 "On", and select the options shown on the right.

(You can choose any of the three marks at the bottom of the screen.)

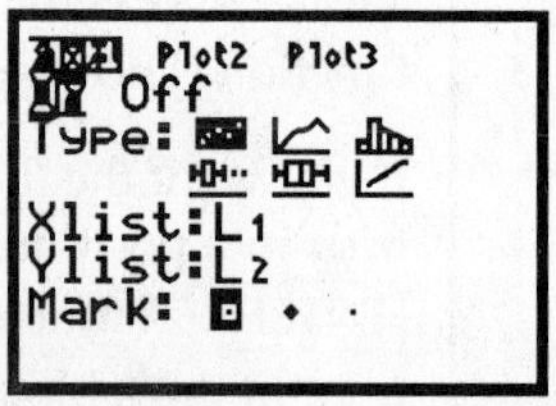

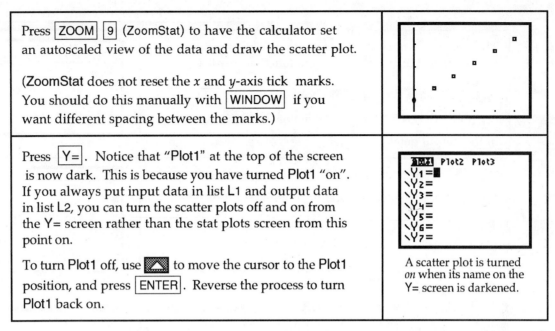

Press ZOOM 9 (ZoomStat) to have the calculator set
an autoscaled view of the data and draw the scatter plot.

(ZoomStat does not reset the x and y-axis tick marks.
You should do this manually with WINDOW if you
want different spacing between the marks.)

Press Y=. Notice that "Plot1" at the top of the screen
is now dark. This is because you have turned Plot1 "on".
If you always put input data in list L1 and output data
in list L2, you can turn the scatter plots off and on from
the Y= screen rather than the stat plots screen from this
point on.

To turn Plot1 off, use ◤◥ to move the cursor to the Plot1
position, and press ENTER. Reverse the process to turn
Plot1 back on.

A scatter plot is turned
on when its name on the
Y= screen is darkened.

- TI-83 lists can be named and stored in the calculator's memory for later recall and use. If
  you do this and use the list by its stored name, you must use the name of the list in the stat
  plot setup or on the stat plot screen each time you change lists. Refer to Section 1.4.13 of
  this *Guide* and your *TI-83 Guidebook* for details.

1.4.6 FINDING FIRST DIFFERENCES   When the input values are evenly spaced, use
program DIFF to compute first differences in the output values. If the data are perfectly
linear (*i.e.*, every data point falls on the graph of the line), the first differences in the
output values are constant. If the first differences are "close" to constant, this is an
indication that a linear model *may* be appropriate.

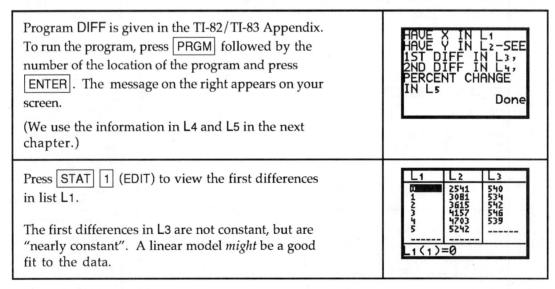

Program DIFF is given in the TI-82/TI-83 Appendix.
To run the program, press PRGM followed by the
number of the location of the program and press
ENTER. The message on the right appears on your
screen.

(We use the information in L4 and L5 in the next
chapter.)

Press STAT 1 (EDIT) to view the first differences
in list L1.

The first differences in L3 are not constant, but are
"nearly constant". A linear model *might* be a good
fit to the data.

- Program DIFF **should not** be used for data with input (L1) values that are *not* evenly
  spaced. First differences give no information about a possible linear fit to data with
  inputs that are not the same distance apart. If you try to use program DIFF with input
  data that are not evenly spaced, the message INPUT VALUES NOT EVENLY SPACED
  appears and the program stops.

TI-83 If you do not want to use program DIFF, you can use your TI-83 to compute first differences of any list. Press 2nd STAT (LIST) ▶ (OPS) 7 (ΔList( ) 2nd 2 (L2)) ENTER to see the list of first differences in the output data.	``` ΔList(L₂) {540 534 542 54… ```
Use ▶ to scroll to the right to see the remainder of the first differences. Use ◀ to scroll back to the left.	``` ΔList(L₂) …34 542 546 539} ▮ ```

1.4.7 FINDING A LINEAR MODEL   Use your calculator to obtain the linear model that best fits the data. Your calculator can find two different, but equivalent, forms of the linear model: $y = ax + b$ or $y = a + bx$. For convenience, we always choose the model $y = ax + b$.

TI-82 Press STAT ▶ (CALC) 5 (LinReg(ax+b)) ENTER. *Note:* If you receive an error message, reset the statistical setup to that indicated on page A-1 of this *Guide*.	``` EDIT CALC 1:1-Var Stats 2:2-Var Stats 3:SetUp… 4:Med-Med 5▮LinReg(ax+b) 6:QuadReg 7↓CubicReg ```
The linear model of best fit for the aligned tax data entered in Section 1.4.4 of this *Guide* is displayed on the home screen. (The *r* that is shown is called the *correlation coefficient*. It is something you will learn about in a statistics course and should be ignored in this course.)	``` LinReg y=ax+b a=540.3714286 b=2538.904762 r=.9999954581 ```
TI-83 Press STAT ▶ (CALC) 4 (LinReg(ax+b)) to copy the instruction to the home screen.	``` EDIT CALC TESTS 1:1-Var Stats 2:2-Var Stats 3:Med-Med 4▮LinReg(ax+b) 5:QuadReg 6:CubicReg 7↓QuartReg ```
To have the calculator find the linear model of best fit using L1 as the input and L2 as the output <u>and</u> paste the model into the Y= list, type the following after the **LinReg(ax+b)** instruction: 2nd 1 (L1) , 2nd 2 (L2) , VARS ▶ (Y-VARS) 1 (Function) 1 (Y1). Press ENTER.	``` LinReg(ax+b) L₁, L₂,Y₁ ``` The model will be pasted into the location that you specify.

The linear model of best fit for the aligned tax data entered in Section 1.4.4 of this *Guide* is displayed on the home screen.

Note: It is not necessary to first clear any previously-entered function from the location of the Y= list.

```
LinReg
 y=ax+b
 a=540.3714286
 b=2538.904762
```

Note: If you see the TI-83 screen shown to the right instead of the screen displayed above, turn off the diagnostics by pressing [2nd] [0] (CATALOG) [x⁻¹], use [▼] to locate DiagnosticOff and press [ENTER] [ENTER].

(The quantities r^2 and r are used in a statistics course and should be ignored in this course.)

```
LinReg
 y=ax+b
 a=540.3714286
 b=2538.904762
 r²=.9999909162
 r=.9999954581
```

Press [Y=] to verify that the model has been pasted into the Y1 location of the graphing list.

Note: If you receive an error message when finding the model, reset the statistical setup to that indicated on page A-1 of this *Guide*.

```
Plot1 Plot2 Plot3
\Y1冒540.37142857
145X+2538.904761
9047
\Y2=
\Y3=
\Y4=
\Y5=
```

1.4.8 PASTING A MODEL INTO THE FUNCTION LIST The coefficients of the model found by the calculator should *not* be rounded. This is not a problem because the calculator will paste the entire model into the function list!

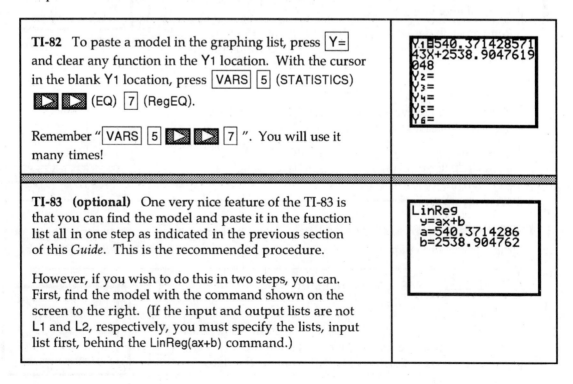

TI-82 To paste a model in the graphing list, press [Y=] and clear any function in the Y1 location. With the cursor in the blank Y1 location, press [VARS] [5] (STATISTICS) [▶] [▶] (EQ) [7] (RegEQ).

Remember "[VARS] [5] [▶] [▶] [7] ". You will use it many times!

```
Y1冒540.371428571
43X+2538.9047619
048
Y2=
Y3=
Y4=
Y5=
Y6=
```

TI-83 (optional) One very nice feature of the TI-83 is that you can find the model and paste it in the function list all in one step as indicated in the previous section of this *Guide*. This is the recommended procedure.

However, if you wish to do this in two steps, you can. First, find the model with the command shown on the screen to the right. (If the input and output lists are not L1 and L2, respectively, you must specify the lists, input list first, behind the LinReg(ax+b) command.)

```
LinReg
 y=ax+b
 a=540.3714286
 b=2538.904762
```

Second, press $\boxed{Y=}$ and clear any function in the Y1 location. With the cursor in the blank Y1 location, press $\boxed{VARS}$ $\boxed{5}$ (STATISTICS) $\boxed{\blacktriangleright}$ $\boxed{\blacktriangleright}$ (EQ) $\boxed{1}$ (RegEQ).

Press $\boxed{Y=}$ and verify that the linear model has been put in Y1.

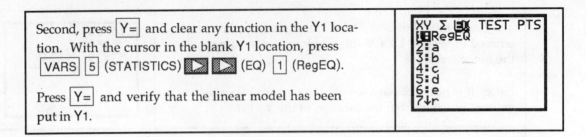

1.4.9 GRAPHING A MODEL

After finding a model, you should always graph it on a scatter plot of the data to verify that the model provides a good fit to the data.

Press $\boxed{GRAPH}$ to overdraw the model on the scatter plot.

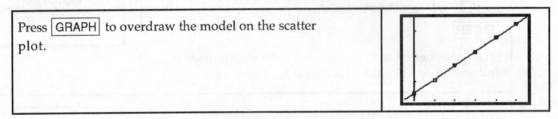

1.4.10 PREDICTIONS USING A MODEL

Use one of the methods described in Sections 1.1.7 or 1.2.1 of this *Guide* to evaluate the linear model at the desired input value. Remember, if you have aligned your data, the input value at which you evaluate the model may not be the value given in the question you are asked.

Predict the tax owed in 1998 and 1999 where the tax is found using the linear model computed from the data given in Section 1.4.1 of this *Guide*:

Tax = $540.37143t + 2538.90476$ dollars and t is the number of years since 1992.

Remember that you should always use the full model, *i.e.*, the function you pasted in Y1, for all computations. Note that 1998 is six years since 1992, so $x = 6$.

The 1998 tax is predicted to be about $5781.

Predict the value of y in 1999 using the **TABLE**. Note that 1999 is seven years since 1992, so $x = 7$. The 1999 tax is predicted to be approximately $6322.

(You can type over x-values already in the table when using Indpnt: ASK, or you can press $\boxed{DEL}$ to delete previously-entered values.)

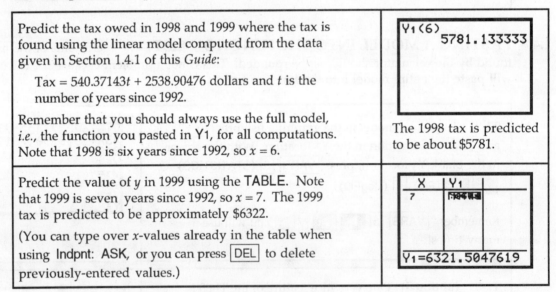

1.4.11 COPYING GRAPHS TO PAPER

Your instructor may ask you to copy what is on your graphics screen to paper. If so, use the following to more accurately perform this task.

TI-82 Press $\boxed{GRAPH}$ to return the graph to the screen.

Press $\boxed{TRACE}$ and $\boxed{\blacktriangleright}$ to trace the graph. The P1 in the upper right-hand corner of the screen indicates that you are tracing the scatter plot of the data. Use either these trace values or the data lists to mark the data points on your paper.

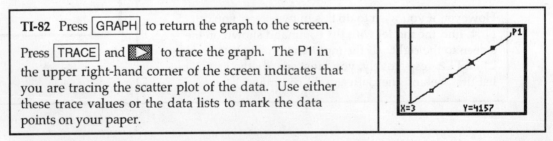

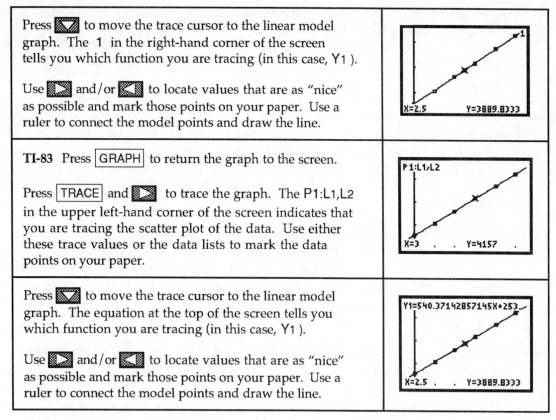

Press ▽ to move the trace cursor to the linear model graph. The 1 in the right-hand corner of the screen tells you which function you are tracing (in this case, Y1).

Use ▷ and/or ◁ to locate values that are as "nice" as possible and mark those points on your paper. Use a ruler to connect the model points and draw the line.

TI-83 Press GRAPH to return the graph to the screen.

Press TRACE and ▷ to trace the graph. The P1:L1,L2 in the upper left-hand corner of the screen indicates that you are tracing the scatter plot of the data. Use either these trace values or the data lists to mark the data points on your paper.

Press ▽ to move the trace cursor to the linear model graph. The equation at the top of the screen tells you which function you are tracing (in this case, Y1).

Use ▷ and/or ◁ to locate values that are as "nice" as possible and mark those points on your paper. Use a ruler to connect the model points and draw the line.

- If you are copying the graph of a continuous curve, rather than a straight line, to your paper, you need to trace as many points as necessary to see the shape of the curve and mark the points on your paper. Connect the points with a smooth curve.

1.4.12 WHAT IS "BEST FIT"? Even though your calculator easily computes the values a and b for the best fitting linear model $y = ax + b$, it is important to understand the method of least-squares and the conditions necessary for its application if you intend to use this model. You can explore the process of finding the line of best fit with program LSLINE. (Program LSLINE is given in the TI-82/TI-83 Appendix.) For your investigations of the least-squares process with this program, it is better to use data that is not perfectly linear and data for which you do *not* know the best-fitting line.

Before using program LSLINE, clear the Y= list and enter your data in lists L1 and L2. Next, draw a scatter plot. Press WINDOW and reset Xscl and Yscl so that you can use the tick marks to help identify points on the graphics screen. Press GRAPH to view the scatter plot.

To activate program LSLINE, press PRGM followed by the number of the location of the program, and press ENTER . The program first displays the scatter plot you constructed and pauses for you to view the screen.

- While the program is calculating, there is a small vertical line in the upper-right hand corner of the graphics screen that is dashed and "moving". The program pauses several times during execution. Whenever this happens, the small vertical line is "still" and you should press ENTER to resume execution after you have looked at the screen.

The program next asks you to find the *y*-intercept and slope of *some* line you estimate will go "through" the data. (You should not expect to guess the best fit line on your first try!)   After you enter a guess for the *y*-intercept and slope, your line is drawn and the errors are shown as vertical line segments on the graph. (You may have to wait just a moment to see the vertical line segments before again pressing ENTER .)

Next, the sum of squares of errors, SSE, is displayed for your line. Choose the TRY AGAIN? option by pressing 1 ENTER . Decide whether you want to move the *y*-intercept of the line or change its slope to improve the fit to the data. After you enter another guess for the *y*-intercept and/or slope, the process of viewing your line, the errors, and display of SSE is repeated. If the new value of SSE is smaller than the SSE for your first guess, you have improved the fit.

When it is felt that an SSE value close to the minimum value is found, you should press 2 at the TRY AGAIN? prompt. The program then overdraws the line of best fit on the graph for comparison with your last attempt and shows the errors for the line of best fit. The coefficients *a* and *b* of the best-fitting linear model $y = ax + b$ are then displayed along with the minimum SSE. Use program LSLINE to explore the method of least squares to find the line of best fit.

1.4.13 NAMING DATA LISTS ON THE TI-83 (optional)   You may or may not want to use the additional features given below for data entered on the TI-83. You can name data (either input, output, or both) and store it in the calculator memory for later recall. For instance, suppose you wanted to name the list L1:   1984   1985   1987   1990   1992

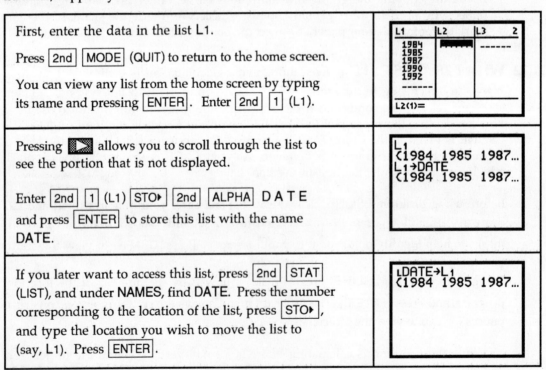

First, enter the data in the list L1. Press 2nd MODE (QUIT) to return to the home screen. You can view any list from the home screen by typing its name and pressing ENTER . Enter 2nd 1 (L1).	L1 L2 L3 2 1984 1985 1987 1990 1992 ------ L2(1)=
Pressing ▶ allows you to scroll through the list to see the portion that is not displayed. Enter 2nd 1 (L1) STO▸ 2nd ALPHA D A T E and press ENTER to store this list with the name DATE.	L1 {1984 1985 1987… L1→DATE {1984 1985 1987…
If you later want to access this list, press 2nd STAT (LIST), and under NAMES, find DATE. Press the number corresponding to the location of the list, press STO▸ , and type the location you wish to move the list to (say, L1). Press ENTER .	LDATE→L1 {1984 1985 1987…

The original data remains in DATE. It is not deleted until you delete it using 2nd + (MEM) 2 (Delete) 4 (List), move the cursor with ▼ to the location of DATE, and press ENTER . Press 2nd MODE (QUIT) to return to the home screen.

Chapter 2 Ingredients of Change: Nonlinear Models

2.1 Exponential Functions and Models

As we consider models that are not linear, it is very important that you be able to use scatter plots, numerical changes in output data, and the underlying shape of the basic functions to be able to identify which model best fits a particular set of data. Finding the model is only a means to an end -- being able to use mathematics to describe the changes that occur in real-world situations.

2.1.1 ENTERING EVENLY-SPACED INPUT VALUES (optional) When an input list consists of many evenly-spaced values, there is a calculator command that will generate the list so that you do not have to type the values in one by one. The syntax for this sequence command is *seq(formula, variable, first value, last value, increment)*. When entering years that differ by 1, the formula is the same as the variable and the increment is 1. Any letter can be used for the variable -- we choose to use *X*.

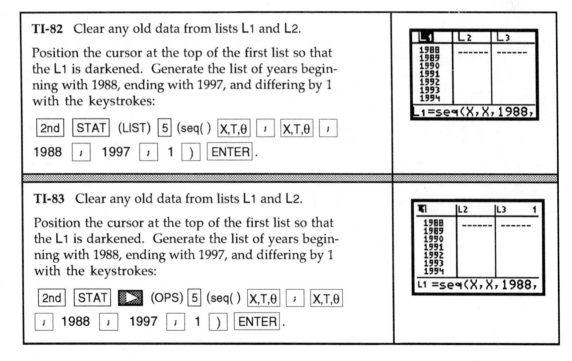

TI-82 Clear any old data from lists L1 and L2.

Position the cursor at the top of the first list so that the L1 is darkened. Generate the list of years beginning with 1988, ending with 1997, and differing by 1 with the keystrokes:

[2nd] [STAT] (LIST) [5] (seq( ) [X,T,θ] [,] [X,T,θ] [,] 1988 [,] 1997 [,] 1 [)] [ENTER].

TI-83 Clear any old data from lists L1 and L2.

Position the cursor at the top of the first list so that the L1 is darkened. Generate the list of years beginning with 1988, ending with 1997, and differing by 1 with the keystrokes:

[2nd] [STAT] [▶] (OPS) [5] (seq( ) [X,T,θ] [,] [X,T,θ] [,] 1988 [,] 1997 [,] 1 [)] [ENTER].

2.1.2 FINDING PERCENTAGE CHANGE When the input values are evenly spaced, use program DIFF to compute percentage change in the output values. If the data are perfectly exponential (i.e., every data point falls on an exponential model), the percentage change in the output values is constant. If the percentage change is "close" to constant, this is an indication that an exponential model *may* be appropriate.

Suppose the population of a small town between the years 1988 and 1997 is as follows:

Year	1988	1989	1990	1991	1992	1993	1994	1995	1996	1997
Population	7290	6707	6170	5677	5223	4805	4420	4067	3741	3442

Clear any old data, and enter the above data in lists L1 (year) and L2 (population). See Section 2.1.1 of this *Guide* for a convenient way to enter the years into L1.

Run program DIFF and press $\boxed{\text{STAT}}$ (1: EDIT) $\boxed{\text{ENTER}}$.

Observe the first differences in L3, the second differences in L4, and the percentage changes in list L5. (Use $\boxed{\blacktriangleright}$ to see L4 and L5.)

The percentage change is very close to constant, so an exponential model may be a good fit.

L3	L4	L5
-493	39	-7.99
-454	36	-7.997
-418	33	-8.003
-385	32	-8.012
-353	27	-7.986
-326	27	-8.016
-299	------	▓▓▓▓

$L_5(9)= \text{-}7.992515...$

Use $\boxed{\blacktriangledown}$ and $\boxed{\blacktriangle}$ to view all of the list.

2.1.3 FINDING AN EXPONENTIAL MODEL   Use your calculator to find an exponential model that fits the data. The exponential model we use is of the form $y = ab^x$.

Construct a scatter plot of the data. Notice that the data curves rather than falling in a straight line pattern. An exponential model certainly seems appropriate!

It is very important that you align large numbers (like years) whenever you find an exponential model. The model found by the calculator may not even be correct if you don't!

Other alignments are possible, but we choose to align so that $x = 0$ in 1988.

L1	L2	L3
0	7290	-583
1	6707	-537
2	6170	-493
3	5677	-454
4	5223	-418
5	4805	-385
6	4420	-353

$L_1 = L_1 - 1988$

TI-82 Fit an exponential model to the data by pressing $\boxed{\text{STAT}}$ $\boxed{\blacktriangleright}$ (CALC) $\boxed{\text{ALPHA}}$ $\boxed{\text{A}}$ (ExpReg) $\boxed{\text{ENTER}}$.

```
EDIT CALC
5↑LinReg(ax+b)
6:QuadReg
7:CubicReg
8:QuartReg
9:LinReg(a+bx)
0:LnReg
A:ExpReg
```

The best-fitting exponential model is displayed on the home screen.

Copy the model to the Y= list (use $\boxed{\text{VARS}}$ $\boxed{5}$ $\boxed{\blacktriangleright}$ $\boxed{\blacktriangleright}$ $\boxed{7}$ as explained in Section 1.4.8 of this *Guide*), overdraw the graph on the scatter plot with $\boxed{\text{ZOOM}}$ $\boxed{9}$ (ZoomStat), and see that it gives a very good fit to the data.

```
ExpReg
y=a*b^x
a=7290.250315
b=.9199949014
r=-.9999999679
```

TI-83 Fit an exponential model to the data and copy the model to the Y1 location of the Y= list by pressing $\boxed{\text{STAT}}$ $\boxed{\blacktriangleright}$ (CALC) $\boxed{0}$ (ExpReg) $\boxed{\text{2nd}}$ $\boxed{1}$ (L1) $\boxed{,}$ $\boxed{\text{2nd}}$ $\boxed{2}$ (L2) $\boxed{,}$ $\boxed{\text{VARS}}$ $\boxed{\blacktriangleright}$ (Y-VARS) $\boxed{1}$ (Function) $\boxed{1}$ (Y1). Press $\boxed{\text{ENTER}}$.

```
ExpReg
y=a*b^x
a=7290.250315
b=.9199949014
```

Both Press GRAPH to overdraw the graph of the model on the scatter plot. The model gives a very good fit to the data.

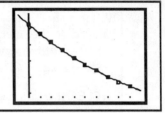

2.1.4 FINDING A LOGISTIC MODEL   Use your calculator to find a logistic model of the

form $y = \dfrac{L}{1 + Ae^{-Bx}}$. The logistic model that you obtain may be slightly different from a logistic model found with another calculator. Logistic models in *Calculus Concepts* were found using a TI-83. Refer to the following TI-82 discussion for the comparable logistic model that best fits the data given in Example 2, Section 2.1 of the text. As with the exponential model $y = ab^x$, large input values must be aligned before fitting a logistic model to data.

(Program LOGISTIC finds a "best-fit" logistic model rather than a logistic model with a user-input limiting value L such that no data value is ever greater than L.)

Clear any old data, and enter the following in lists L1 and L2:

Aligned end of month	1	2	3	4	5	6	7	8	9
Total swimsuits sold	4	12	25	58	230	439	648	748	769

Construct a scatter plot of the data. A logistic model seems appropriate.	
TI-82 Use program[1] LOGISTIC to fit the logistic model. **Note:** To use this program, the input data must be in order, from smallest to largest, in list L1. Have the corresponding output data in L2. Run program LOGISTIC with PRGM followed by the number of the location of the program. Press ENTER .	EXEC EDIT NEW 1:AUTOSCL 2:DIFF 3:EULER 4:LOGISTIC 5:LSLINE 6:NUMINTGL 7↓SECTAN Your program list may not look exactly like this.
The first message you see reminds you that the input data should be in list L1 and the output data in list L2 . If you have not done this, press ON and choose 2 (QUIT). Enter the data and then rerun the program.	DATA IN L₁,L₂ ENTER CONTINUES
After pressing ENTER to continue, the program displays several messages to let you know it is working. (These messages can be ignored. You can see *SSE* being reduced as the model is fit.)	STEP: 1 WORKING... SSE: 11093.42805

---

[1]The authors express their sincere appreciation to Dr. Dan Warner and Robert Simms of the Mathematical Sciences Department at Clemson University for their invaluable help with program LOGISTIC .

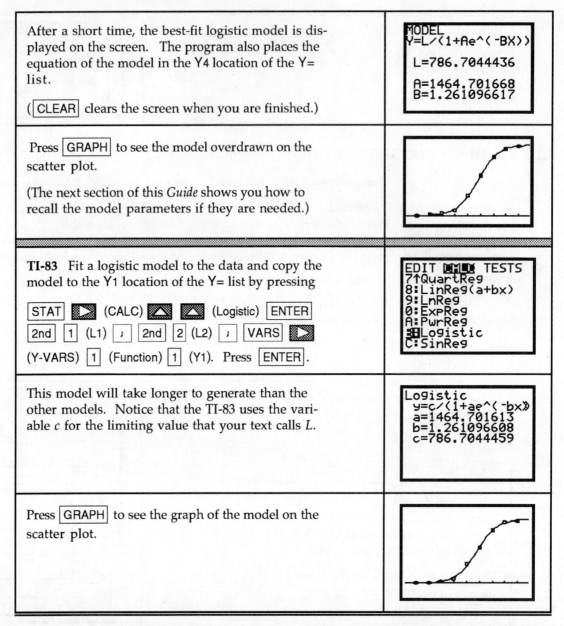

After a short time, the best-fit logistic model is displayed on the screen. The program also places the equation of the model in the Y4 location of the Y= list.

([CLEAR] clears the screen when you are finished.)

```
MODEL
Y=L/(1+Ae^(-BX))

L=786.7044436

A=1464.701668
B=1.261096617
```

Press [GRAPH] to see the model overdrawn on the scatter plot.

(The next section of this *Guide* shows you how to recall the model parameters if they are needed.)

TI-83 Fit a logistic model to the data and copy the model to the Y1 location of the Y= list by pressing

[STAT] [▷] (CALC) [▲] [▲] (Logistic) [ENTER]
[2nd] [1] (L1) [,] [2nd] [2] (L2) [,] [VARS] [▷]
(Y-VARS) [1] (Function) [1] (Y1). Press [ENTER].

```
EDIT CALC TESTS
7↑QuartReg
8:LinReg(a+bx)
9:LnReg
0:ExpReg
A:PwrReg
B:Logistic
C:SinReg
```

This model will take longer to generate than the other models. Notice that the TI-83 uses the variable *c* for the limiting value that your text calls *L*.

```
Logistic
  y=c/(1+ae^(-bx))
  a=1464.701613
  b=1.261096608
  c=786.7044459
```

Press [GRAPH] to see the graph of the model on the scatter plot.

- **Both** Provided the input values are evenly spaced, program DIFF might be helpful when you are trying to determine if a logistic model is appropriate for certain data. If the first differences (in list L3 after running program DIFF) *begin small*, *peak in the middle*, and *end small*, this is an indication that a logistic model may provide a good fit to the data. Such is true for this data set because the first differences are 8, 13, 33, 172, 209, 209, 100, and 21.

2.1.5 RECALLING MODEL PARAMETERS   Rounding of model parameters can often lead to incorrect or misleading results. You may find that you need to use the full values of model parameters after you have found a model.

It would be tedious to copy all these digits into another location of your calculator. You don't have to! The following procedure applies for any model you find using one of the built-in regressions (*i.e.*, from the STAT CALC menu) in your calculator. Of course, once another model is found, previous parameters are no longer stored in the calculator's memory.

TI-82 As an example, we use the parameter b for the exponential model found in Section 2.1.3. However, this same procedure applies to any model you have found using the STAT CALC menu. To recall the value of b in the model $y = ab^x$, press [VARS] [5] (Statistics) [▷] [▷] (EQ) [2] (b) [ENTER] .	X/Y Σ EQ BOX PTS 1:a 2:b 3:c 4:d 5:e 6:r 7:RegEQ
The full value of b will be "pasted" where ever you had the cursor before beginning the above keystrokes. Note: Remember that you can only recall the parameters before you use the menu to find a different model. Once a different model is found, the parameters are given for the new model.	a=7290.250315 b=.9199949014 r=-.9999999679 ln b -.0833871509

The above procedure does not apply to models found with programs you have entered in the calculator.
 	
If you need to recall the values of A, B, or L for the logistic model, return to the home screen and type [ALPHA] [MATH] (A) [ENTER] , [ALPHA] [MATRIX] (B) [ENTER] , and then [ALPHA] [ )] (L) [ENTER] .
	A
	      1464.701668
	B
	      1.261096617
	L
	      786.7044436

TI-83 As an example, we use the parameter c for the logistic model found in Section 2.1.4. However, this same procedure applies to any model you have found using the STAT CALC menu. To recall the value of c in $y = c/(1+ae^{\wedge}(\bar{\ }bx))$, press [VARS] [5] (Statistics) [▷] [▷] (EQ) [4] (c) [ENTER] .	XY Σ EQ TEST PTS 1:RegEQ 2:a 3:b 4:c 5:d 6:e 7↓r
The full value of c will be "pasted" where ever you had the cursor before beginning the above keystrokes.	L1 L2 L3 3 1 4 ┌──┐ 2 12 ────── 3 25 4 58 5 230 6 439 7 648 L3(1)=786.7044459...

2.1.6 RANDOM NUMBERS

Imagine all the real numbers between 0 and 1, including the 0 but not the 1, written on identical slips of paper and placed in a hat. Close your eyes and draw one slip of paper from the hat. You have just chosen a number "at random."

Your calculator doesn't offer you a random choice of all real numbers between 0 and 1, but it allows you to choose, *with an equal chance of obtaining each one*, any of 10^{14} different numbers between 0 and 1 with its random number generator called rand.

First, "seed" the random number generator. (This is like mixing up all the slips of paper in the hat.)

Pick some number, <u>not</u> the one shown on the right, and store it as the "seed". (Everyone needs to have a different seed, or the choice will not be random.) The random number generator is accessed with MATH ◀ (PRB) 1 (rand).	```658→rand``` 658
Enter **rand** again, and press ENTER several times. Your list of random numbers should be different from the one on the right if you entered a different seed.	rand .5789882558 .1825542239 .286871645 .7527620949 .8421450981 .8800382855

If you want to choose, at random, a whole number between 1 and N, enter int(N rand + 1) by pressing MATH ▶ 4 (int) [(N MATH ◀ (PRB) 1 (rand) + 1)] ENTER f or a specific value of N.
 	
Repeatedly press ENTER to choose more random numbers. For instance, the screen to the right shows several values that were chosen with $N = 10$.
	int (10rand+1)
	                    4
	                    2
	                    4
	                    7
	                    7
	                    1

2.2 Exponential Models in Finance

You are probably familiar with the compound interest formulas. This section introduces you to some new methods of using your calculator with familiar formulas.

2.2.1 REPLAY OF PREVIOUS ENTRIES TO FIND FORMULA OUTPUTS   You can recall expressions previously typed by repeatedly using the calculator's last entry feature. Learn to use this time-saving feature of your calculator.

On the home screen, type the formula $\left(1+\frac{1}{n}\right)^n$ and press ENTER. The output depends on the value of n. You probably obtained a different output value because you have a different value stored in N.
 	
Store 1 in N.
	(1+1/N)^N
	          2.691588029
	1→N
	                    1

Press 2nd ENTER (ENTRY) twice (or as many times as it takes to again display the amount formula on the screen), and then press ENTER. The formula is now evaluated at N = 1.
 	
Store 2 in N and repeat the procedure.
	(1+1/N)^N 1
	                    2
	2→N
	                    2
	(1+1/N)^N
	                2.25

Because this formula contains only one input variable, you could enter it in the Y= list, using X as the input variable, and find the outputs using the TABLE.

Enter Y1 = (1 + 1/X)^X.

Refer to 1.2.1 of this *Guide* to review the information about evaluating outputs using the TABLE.

X	Y1	
1	2	
2	2.25	
X=		

2.2.2 DETERMINING FUTURE VALUE   You can save a lot of keystrokes by recalling expressions previously typed by repeatedly using the calculator's last entry feature. When a formula contains more than one input variable, it is easier to recall the last entry on the home screen than to try to use the TABLE. To illustrate, consider the compound interest formula -- one that contains several input variables.

On the home screen, type in the formula for the amount in an account paying $r\%$ interest (compounded n times a year) on an initial deposit of $\$P$ over a period of t years:

$$Amount = P\left(1 + \frac{r}{n}\right)^{nt} \text{ dollars}$$

The value you obtain with [ENTER] depends on the values your calculator has stored in P, R, and T.

```
1000→P
              1000
.05→R
               .05
4→N
                 4
1→T
```
Store 1000 in P, 0.05 in R, 4 in N, and 1 in T.

Press [2nd] [ENTER] (ENTRY) three times (or as many times as it takes to again display the formula on the screen) and then press [ENTER]. The formula has been evaluated at the stored values of the variables.

```
              1000
.05→R
               .05
4→N
                 4
P(1+R/N)^(NT)
       1050.945337
```

Determine the accumulated amount in an account if $5000 is invested at 5% interest compounded monthly for 3 years by repeating the procedure described above.

(Note that since the value of R has not changed, it is not necessary to again store 0.05 to R.)

The 3-year future value of the amount in the account is $5807.36.

```
5000→P
              5000
12→N:3→T
                 3
P(1+R/N)^(NT)
       5807.361157
```

2.2.3 FINDING PRESENT VALUE   The present value of an investment is easily found with the calculator's solve routine. For instance, suppose you want to solve the equation $9438.40 = P\left(1 + \frac{0.075}{12}\right)^{60}$ for the present value P.

TI-82   Refer to Section 1.2.2 of this *Guide* for instructions on using the TI-82's solve routine and enter the expression on the right.

Remember that a *guess*, here entered as 7000, can be obtained from viewing a graph of Y1 = 9438.4 − X(1+.075/12)^60 and tracing to the approximate location where the graph crosses the horizontal axis.

```
solve(9438.4−P(1
+.075/12)^60,P,7
000)
         6494.48587
```

TI-83 For this illustration, we use solver Method 2 discussed in Section 1.2.2 of this *Guide*.

Press MATH 0 to access the solver, ▲ CLEAR to clear any previously-entered equation, and enter the expression on the right.

```
EQUATION SOLVER
eqn:0=9348.4-P(1
+.075/12)^60
```

Press ENTER . With the blinking cursor on the line with *P*, press ALPHA ENTER to find *P* = $6432.56.

Remember that a *guess* for *P* should be entered before solving if there is more than one solution. The number of possible solutions as well as guesses for their values can be obtained from viewing a graph of Y1 = 9438.4 − X(1 + 0.075/12)^60 and tracing to the approximate locations where the graph crosses the horizontal axis.

```
9348.4-P(1+.0...=0
■P=6432.5576061...
 bound=(-1E99, 1...
■left-rt=0
```

- If you prefer, you could find the present value by graphically locating the *x*-intercept of Y1 = 9438.4 − X(1 + 0.075/12)^60. Refer to Section 1.2.3 of this *Guide* for more detailed instructions.

2.3 Polynomial Functions and Models

You will in this section learn how to fit models to data that have the familiar shape of a parabola or a cubic. Using your calculator to find these models involves basically the same procedure as when using it to find linear and exponential models.

2.3.1 FINDING SECOND DIFFERENCES When the input values are evenly spaced, use program DIFF to compute second differences in the output values. If the data are perfectly quadratic (*i.e.*, every data point falls on a quadratic model), the second differences in the output values are constant. If the second differences are "close" to constant, this is an indication that a quadratic model *may* be appropriate.

Clear any old data, and enter the roofing job data in lists L1 and L2:

Months after January	1	2	3	4	5	6
Number of jobs	12	14	22	37	58	84

The input values are evenly spaced, so we can see what information is given by viewing the second differences.

Run program DIFF and observe the first differences in list L3, the second differences in L4, and the percentage differences (changes) in list L5.

The second differences are close to constant, so a quadratic model may be a good fit.

Construct a scatter plot of the data. A quadratic model seems appropriate!

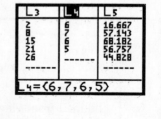

2.3.2 FINDING A QUADRATIC MODEL Use your calculator to obtain a quadratic model that fits the data. Your calculator's quadratic model is of the form $y = ax^2 + bx + c$.

TI-82 Press STAT ▶ (CALC) 6 (QuadReg) ENTER .	EDIT CALC 1:1-Var Stats 2:2-Var Stats 3:SetUp... 4:Med-Med 5:LinReg(ax+b) 6▉QuadReg 7↓CubicReg
The best-fitting quadratic model is displayed on the home screen.	QuadReg y=ax²+bx+c a=3.071428571 b=-7.014285714 c=15.8
Copy the model to the Y= list, overdraw the graph on the scatter plot, and see that it gives a good fit to the data.	
TI-83 Fit a quadratic model and copy it to the graphing list by pressing STAT ▶ (CALC) 5 (QuadReg) 2nd 1 (L1) , 2nd 2 (L2) , VARS ▶ (Y-VARS) 1 (Function) 1 (Y1).	QuadReg L₁,L₂,Y₁
Press ENTER to see the model. Overdraw the graph of the model on the scatter plot.	QuadReg y=ax²+bx+c a=3.071428571 b=-7.014285714 c=15.8

2.3.3 FINDING A CUBIC MODEL   Whenever a scatter plot of the data shows a single change in concavity, a cubic or logistic model is appropriate. If a limiting value is apparent, use the logistic model. Otherwise, a cubic model should be considered. When appropriate, use your calculator to obtain the cubic model that best fits data. The calculator's cubic model is of the form $y = ax^3 + bx^2 + cx + d$.

Clear any old data, and enter the average price in dollars per 1000 cubic feet of natural gas for residential use in the U.S. from 1980 through 1990 in lists L1 and L2:

Year	1980	1981	1982	1983	1984	1985	1986	1987	1988	1989	1990
Price	3.68	4.29	5.17	6.06	6.12	6.12	5.83	5.54	5.47	5.64	5.77

First, clear your lists, and then enter the data.

In order to work with smaller coefficients, align the data so that x represents the number of years since 1980.

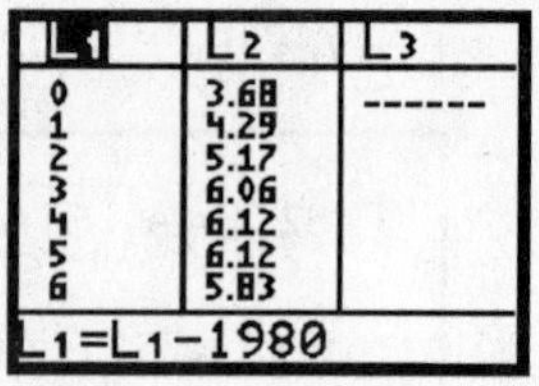

L₁	L₂	L₃
0	3.68	------
1	4.29	
2	5.17	
3	6.06	
4	6.12	
5	6.12	
6	5.83	

$L_1 = L_1 - 1980$

Draw a scatter plot of these data.

Notice that a concavity change is evident, but there do not appear to be any limiting values. Thus, a cubic model is appropriate to fit the data.

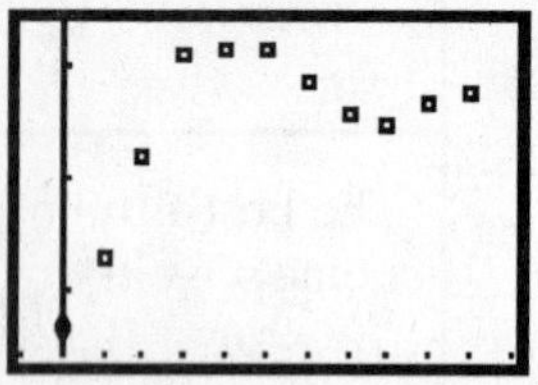

TI-82 Press [STAT] [▶] (CALC) [7] (CubicReg) [ENTER].

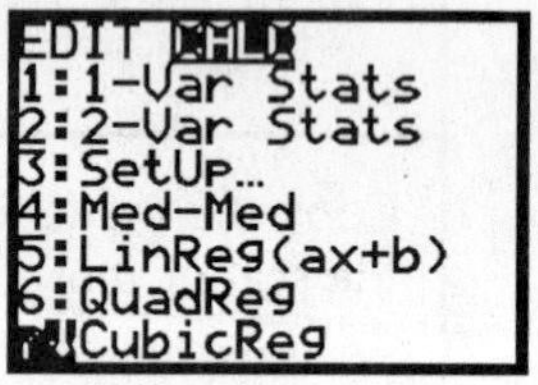

EDIT CALC
1:1-Var Stats
2:2-Var Stats
3:SetUp…
4:Med-Med
5:LinReg(ax+b)
6:QuadReg
7:CubicReg

The best-fitting cubic model is displayed on the home screen.

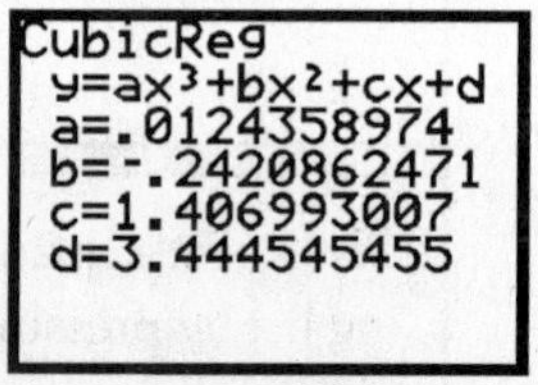

CubicReg
  y=ax³+bx²+cx+d
  a=.0124358974
  b=-.2420862471
  c=1.406993007
  d=3.444545455

Copy the model to the Y= list, overdraw the graph on the scatter plot, and see that it gives a reasonably good fit to the data.

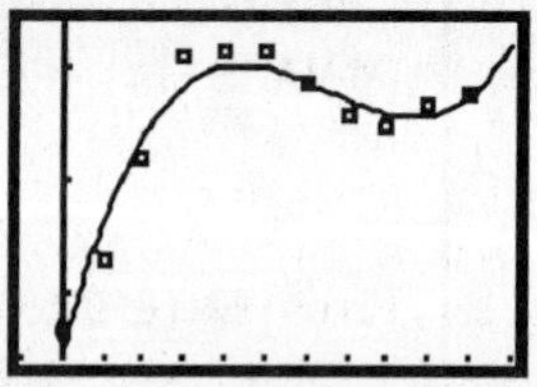

TI-83 Fit a cubic model and copy it to the graphing list by pressing [STAT] [▶] (CALC) [6] (CubicReg) [2nd] [1] (L1) [,] [2nd] [2] (L2) [,] [VARS] [▶] (Y-VARS) [1] (Function) [1] (Y1).

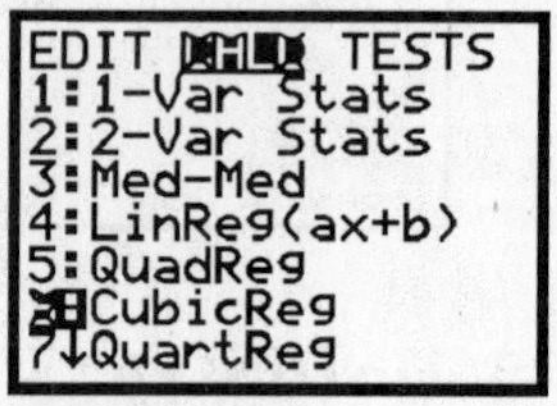

EDIT CALC TESTS
1:1-Var Stats
2:2-Var Stats
3:Med-Med
4:LinReg(ax+b)
5:QuadReg
6:CubicReg
7↓QuartReg

Press [ENTER] to see the model and to overdraw the graph of the model on the scatter plot.

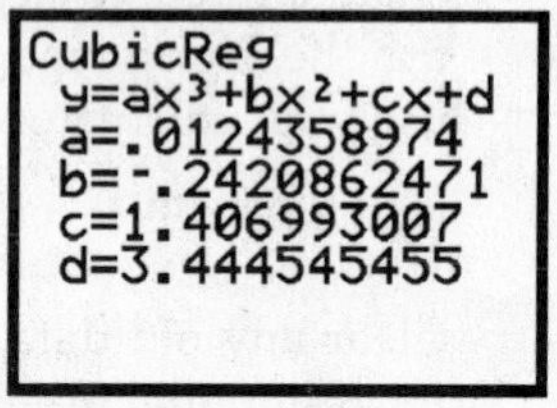

CubicReg
  y=ax³+bx²+cx+d
  a=.0124358974
  b=-.2420862471
  c=1.406993007
  d=3.444545455

Chapter 3 Describing Change: Rates

3.1 Average Rates of Change

As you calculate average and other rates of change, remember that each numerical answer should be accompanied by units telling how the quantity is measured. You should also be able to interpret each numerical answer. It is only through their interpretations that the results of your calculations will be useful in real-world situations.

3.1.1 FINDING AVERAGE RATES OF CHANGE   Finding an average rate of change using a model is just a matter of evaluating the model at two different values of the input variable and dividing by the difference in those input values. Consider this example.

The population density of Nevada from 1950 through 1990 can be approximated by the model $P(t) = 0.1273(1.05136)^t$ people per square mile where t is the number of years since 1900. You are asked to calculate the average rates of change between from 1950 through 1980 and between 1980 and 1990.

Enter the equation in the Y1 location of the Y= list. (Remember that you must use x as the input variable in the graphing list. You do not have to use the first function location -- any of them will do.)	```Y₁■.1273(1.05136)^X Y₂= Y₃= Y₄= Y₅= Y₆= Y₇=```
Return to the home screen with 2nd MODE (QUIT). The average rate of change of the population density between 1950 and 1980 is $\dfrac{P(80) - P(50)}{80 - 50} = \dfrac{Y_1(80) - Y_1(50)}{80 - 50}$. Enter this quotient, remembering to use parentheses around both the numerator and the denominator.	```(Y₁(80)-Y₁(50))/ (80-50) .1813373771```
To find the average rate of change between 1980 and 1990, simply recall the last expression with 2nd ENTER (ENTRY) and replace the 50 by 90. Press ENTER.	```(Y₁(80)-Y₁(50))/ (80-50) .1813373771 (Y₁(80)-Y₁(90))/ (80-90) .4549202871```

Recall that rate of change units are output units per input units. We see that on average, the population density increased by about 0.18 person per square mile per year between 1950 and 1980 and by approximately 0.45 person per square mile per year between 1980 and 1990.

- If you have many average rates of change to calculate, you could put the average rate of change formula in the graphing list: Y2 = (Y1(A) – Y1(B))/(A – B). (You, of course, need to have a model in Y1.) Then, on the home screen, store the inputs of the two points in A and B: 80 → A : 90 → B. All you need do then is type Y1 and press enter. Store the next set of inputs into A and B and use 2nd ENTER to recall Y1 to find the average rate of change between the two new points.   Try it!

📖 3.3 Tangent Lines

We first examine the principle of local linearity which says that if you are close enough, the tangent line and the curve are indistinguishable. We then use the calculator to draw tangent lines. There are two ways you can have your calculator draw a tangent line at a point on a curve. In this section, we consider one of these. The other method will be discussed in Chapter 4 of this *Guide*.

3.3.1 MAGNIFYING A PORTION OF A GRAPH   The ZOOM menu of your calculator allows you to magnify any portion of the graph of a function. Suppose we are investigating the graph of $y = ^-x^2 + 40x +50$ and the tangent line, $y = 20x + 150$, to the graph of this function at $x = 10$.

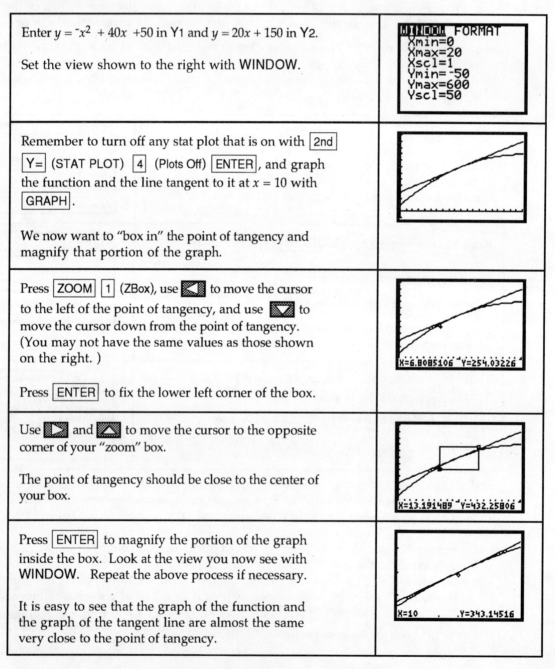

Enter $y = ^-x^2 + 40x +50$ in Y1 and $y = 20x + 150$ in Y2. Set the view shown to the right with WINDOW.	WINDOW FORMAT Xmin=0 Xmax=20 Xscl=1 Ymin=-50 Ymax=600 Yscl=50
Remember to turn off any stat plot that is on with 2nd Y= (STAT PLOT) 4 (Plots Off) ENTER , and graph the function and the line tangent to it at $x = 10$ with GRAPH . We now want to "box in" the point of tangency and magnify that portion of the graph.	
Press ZOOM 1 (ZBox), use ◀ to move the cursor to the left of the point of tangency, and use ▼ to move the cursor down from the point of tangency. (You may not have the same values as those shown on the right.) Press ENTER to fix the lower left corner of the box.	X=6.8085106 Y=254.03226
Use ▶ and ▲ to move the cursor to the opposite corner of your "zoom" box. The point of tangency should be close to the center of your box.	X=13.191489 Y=432.25806
Press ENTER to magnify the portion of the graph inside the box. Look at the view you now see with WINDOW. Repeat the above process if necessary. It is easy to see that the graph of the function and the graph of the tangent line are almost the same very close to the point of tangency.	X=10 .Y=343.14516

3.3.2 DRAWING A TANGENT LINE   The DRAW menu of your calculator contains the instruction to draw a tangent line to a curve at a point. To illustrate the process, we draw several tangent lines on the graph of $f(x) = x^3 + x^2 - 10x - 2$. We also investigate what the calculator does when you ask it to draw a tangent line where the line cannot be drawn.

Enter $f(x) = x^3 + x^2 - 10x - 2$ in Y1. Set the view shown to the right with **WINDOW**. (The TI-83 window does not have "format" at the top.) Press GRAPH.	```WINDOW FORMAT Xmin=-4.7 Xmax=4.7 Xscl=1 Ymin=-30 Ymax=30 Yscl=10```
TI-82 Return to the home screen with 2nd MODE (QUIT). Draw the tangent line to the curve at $x = 0$ with 2nd PRGM (DRAW) 5 (Tangent( ) 2nd VARS (Y-VARS) 1 (Function) 1 (Y1) , 0) ENTER.	```Tangent(Y₁,0)```
TI-83 Return to the home screen with 2nd MODE (QUIT). Draw the tangent line to the curve at $x = 0$ with 2nd PRGM (DRAW) 5 (Tangent( ) VARS ▶ (Y-VARS) 1 (Function) 1 (Y1) , 0) ENTER.	```Tangent(Y₁,0)```
Both Notice that the tangent line cuts through the curve at $x = 0$. It appears that $(0, {}^-2)$ is an inflection point.	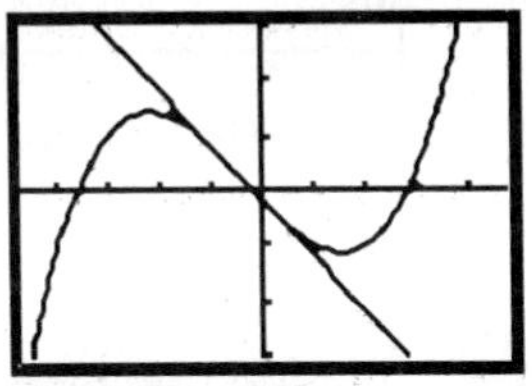
Return to the home screen, and recall the last entry with 2nd ENTER (ENTRY). Edit the statement so that you can draw the tangent line at $x = {}^-3$.	```Tangent(Y₁,0) Tangent(Y₁,-3)```
Once again recall the last entry on the home screen, and then draw the tangent line at $x = 1.5$. The tangent line is almost, but not quite, horizontal at $x = 1.5$.	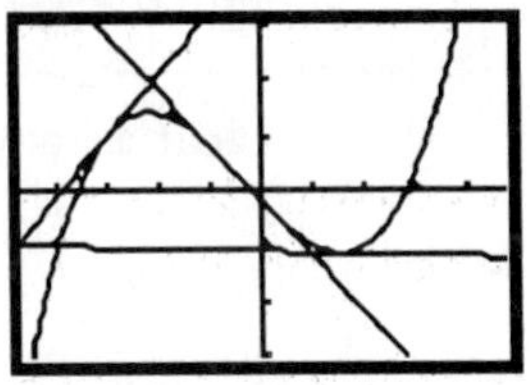

Let us now look at some special cases:

1. What happens if the tangent line is vertical? We consider the function $f(x) = (x + 1)^{1/3}$ which has a vertical tangent at $x = {}^-1$.

2. How does the calculator respond when the tangent line cannot be drawn at a point? We illustrate what happens with $g(x) = |x| - 1$, a function that has a sharp point at $(0, {}^-1)$.

3. Does the calculator draw the tangent line at the joining point(s) of a piecewise continuous function? We consider two situations:

   a. $h(x)$, a piecewise continuous function that is continuous at all points and

   b. $m(x)$, a piecewise continuous function that is not continuous at $x = 1$.

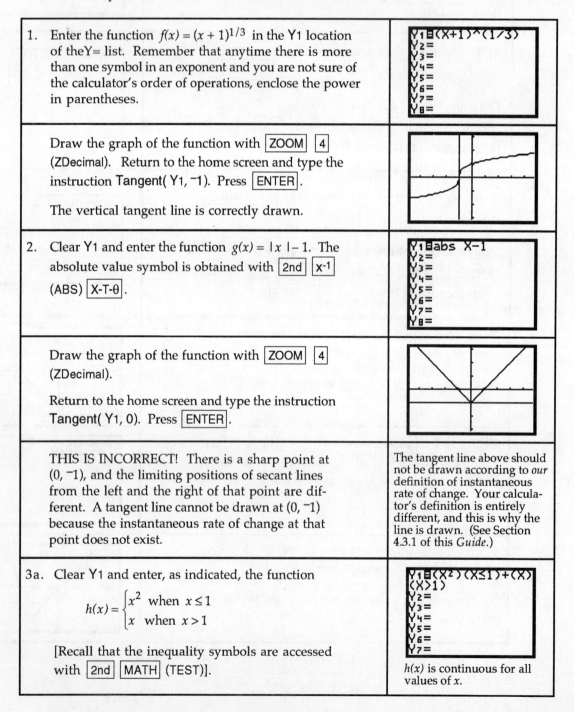

1. Enter the function $f(x) = (x + 1)^{1/3}$ in the Y1 location of the Y= list. Remember that anytime there is more than one symbol in an exponent and you are not sure of the calculator's order of operations, enclose the power in parentheses.	`Y₁█(X+1)^(1/3)`
	`Y₂=`
	`Y₃=`
	`Y₄=`
	`Y₅=`
	`Y₆=`
	`Y₇=`
	`Y₈=`

Draw the graph of the function with ZOOM 4 (ZDecimal). Return to the home screen and type the instruction Tangent(Y1, ⁻1). Press ENTER. The vertical tangent line is correctly drawn.	

2. Clear Y1 and enter the function $g(x) = |x| - 1$. The absolute value symbol is obtained with 2nd x⁻¹ (ABS) X-T-θ.	`Y₁█abs X-1`
	`Y₂=`
	`Y₃=`
	`Y₄=`
	`Y₅=`
	`Y₆=`
	`Y₇=`
	`Y₈=`

Draw the graph of the function with ZOOM 4 (ZDecimal). Return to the home screen and type the instruction Tangent(Y1, 0). Press ENTER.	
THIS IS INCORRECT! There is a sharp point at $(0, {}^-1)$, and the limiting positions of secant lines from the left and the right of that point are different. A tangent line cannot be drawn at $(0, {}^-1)$ because the instantaneous rate of change at that point does not exist.	The tangent line above should not be drawn according to *our* definition of instantaneous rate of change. Your calculator's definition is entirely different, and this is why the line is drawn. (See Section 4.3.1 of this *Guide*.)

3a. Clear Y1 and enter, as indicated, the function
 	
$$h(x) = \begin{cases} x^2 & \text{when } x \le 1 \\ x & \text{when } x > 1 \end{cases}$$
 	
[Recall that the inequality symbols are accessed with 2nd MATH (TEST)].
	`Y₁█(X²)(X≤1)+(X)`
	`(X>1)`
	`Y₂=`
	`Y₃=`
	`Y₄=`
	`Y₅=`
	`Y₆=`
	`Y₇=`
	
	$h(x)$ is continuous for all values of x.

Draw the graph of the function with [ZOOM] [4] (ZDecimal).

Return to the home screen and enter Tangent(Y1, 1).

THE GRAPH YOU SEE IS INCORRECT because secant lines drawn with points on the right and left of $x = 1$ do not approach the same slope.

The tangent line above should *not* be drawn.

3b. Edit Y1 to enter, as indicated, the function

$$m(x) = \begin{cases} x^2 & \text{when } x \le 1 \\ x+1 & \text{when } x > 1 \end{cases}$$

TI-82 Press [MODE] and choose Dot.

TI-83 Move the cursor to the left of Y1 and press [ENTER] six times to change the slanted line to a dotted line.

Both Draw the graph of the function with [ZOOM] [4] (ZDecimal).

Since $m(x)$ is not continuous at $x = 1$, the instantaneous rate of change does not exist at that point. The tangent line cannot be drawn at (1, 1).

TI-82 Go to the MODE screen and return your calculator to Connected mode.

TI-83 Press [Y=], use ◄ to move to the left of Y1, and press [ENTER] once to return your calculator to connected mode.

Both On the home screen, type the instruction Tangent(Y1, 1). Press [ENTER].

THE GRAPH YOU SEE IS INCORRECT. A vertical tangent is drawn, but the tangent line certainly does not exist when $x = 1$.

Caution: Be certain that the instantaneous rate of change exists at a point before using your calculator to draw a tangent line at that point. Because of the way your calculator computes instantaneous rates of change, it may draw a tangent line at a point on a curve where the tangent line does not exist.

📖 3.5 Percentage Change and Percentage Rates of Change

The calculations in this section involve no new calculator techniques. When calculating percentage change or percentage rates of change, you have the option of using a program or the home screen.

3.5.1 CALCULATING PERCENTAGE CHANGE

Recall that program DIFF stores percentage changes (also called percentage differences) in output data in list L5. Consider the following data giving quarterly earnings for a business:

Quarter ending	Mar 1994	June 1994	Sept 1994	Dec 1994	Mar 1995	June 1995
Earnings (millions)	27.3	28.9	24.6	32.1	29.4	27.7

First, we enter the data in the calculator's lists L1 and L2.

Align the input data so that x is the number of quarters since March 1994. Input x in L1 and earnings (in millions) in L2.	L1: 0,1,2,3,4,5 L2: 27.3, 28.9, 24.6, 32.1, 29.4, 27.7 L2(7)=
Run program DIFF and view the percentage change in list L5. Notice that the percentage change from the end of September 1994 through December 1994 is about 30.5%. Also, from the end of March 1995 through June 1995, the percentage change is approximately ‾5.8%.	L3: 1.6, ‾4.3, 7.5, ‾2.7, ‾1.7 L4: ‾5.9, 11.8, ‾10.2, 1 L5: 5.8608, ‾14.88, 30.488, ‾8.411, ‾5.782 L5=(5.860805860...
You may find it easier to calculate these values using the percentage change formula than have program DIFF do it for you.	(32.1-24.6)/24.6 *100 30.48780488 (27.7-29.4)/29.4 *100 -5.782312925

3.5.2 CALCULATING PERCENTAGE RATE OF CHANGE

Consider again the quarterly earnings for a business. Suppose you are told or otherwise find that the rate of change at the end of the June 1994 is 1.8 million dollars per quarter. Evaluate the percentage rate of change at the end of June 1994.

Divide the rate of change at the end of June 1994 by the earnings, in millions, at the end of June 1994 and multiply by 100 to obtain the percentage rate of change at that point. The percentage rate of change in earnings at the end of June 1994 was approximately 6.2% per quarter.	1.8/28.9 .062283737 Ans*100 6.228373702

Chapter 4   Determining Change: Derivatives

📖 4.1 Numerically Finding Slopes

Using your calculator to find slopes of tangent lines does not involve a new procedure. However, the techniques in this section allow you to repeatedly apply a method of finding slopes that gives quick and accurate results.

4.1.1 NUMERICALLY INVESTIGATING SLOPES ON THE HOME SCREEN

Finding slopes of secant lines joining the point at which the tangent line is drawn to increasingly close points on a function to the left and right of the point of tangency is easily done using your calculator. Suppose we want to find the slope of the tangent line at $t = 8$ to the graph of the function giving the number of polio cases in 1949: $y = \dfrac{42183.911}{1 + 21484.253e^{-1.248911t}}$

where $t = 1$ on January 31, 1949, $t = 2$ on February 28, 1949, and so forth.

Enter the equation in the Y1 location of the Y= list. (Carefully check the entry of your equation, especially the location of the parentheses.) We now evaluate the slopes joining nearby points to the *left* of $x = 8$.	```Y1=42183.911/(1+ 21484.253e^(-1.2 48911X)) Y2= Y3= Y4= Y5= Y6=```
Type in the expression shown to the right to compute the slope of the secant line joining $x = 7.9$ and $x = 8$. You must use parentheses around both the numerator and the denominator of the slope formula. Record each slope on your paper as it is computed. You are trying to find what these slopes are approaching.	```(Y1(7.9)-Y1(8))/ (7.9-8) 13159.68272```
Press ⌈2nd⌉ ⌈ENTER⌉ (ENTRY) to recall the last entry, and then use the cursor keys to move the cursor over the 9 in the "7.9". Press ⌈2nd⌉ ⌈DEL⌉ (INS) and press ⌈9⌉ to insert another 9 in <u>both</u> positions where 7.9 appears. Press ⌈ENTER⌉ to find the slope of the secant line joining $x = 7.99$ and $x = 8$.	```(Y1(7.9)-Y1(8))/ (7.9-8) 13159.68272 (Y1(7.99)-Y1(8)) /(7.99-8) 13170.61766```
Continue in this manner, recording each result, until you can determine to which value the slopes from the left seem to be getting closer and closer.	``` 13170.61766 (Y1(7.999)-Y1(8))/(7.999-8) 13170.18713 (Y1(7.9999)-Y1(8))/(7.9999-8) 13170.12882```
We now evaluate the slopes joining nearby close points to the *right* of $x = 8$. Clear the screen with ⌈CLEAR⌉, recall the last expression with ⌈2nd⌉ ⌈ENTER⌉ (ENTRY), and edit it with ⌈DEL⌉ so that the nearby point is $x = 8.1$. Press ⌈ENTER⌉.	```(Y1(8.1)-Y1(8))/ (8.1-8) 13146.38421```

Continue in this manner as before, recording each result on paper, until you can determine the value the slopes from the right seem to be approaching.

When the slopes from the left and the slopes from the right approach the same value, that value is the slope of the tangent line at $x = 8$.

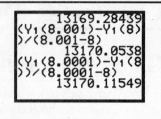

The slopes from the left and from the right appear to be getting closer and closer to 13,170. (The number of polio cases makes sense only as a whole number.)

4.1.2 NUMERICALLY INVESTIGATING SLOPES USING THE TABLE

The process shown in Section 4.1.1 can be done in fewer steps when you use the TABLE. Recall that we are evaluating the slope formula

$$\frac{f(x + h) - f(x)}{(x + h) - x} = \frac{f(8 + h) - f(8)}{h}$$

for various values of h where h is the distance from 8 to the input of the close point. This process is illustrated using the logistic function given in Section 4.1.1 of this *Guide*.

Remember that when in the graphing list, you must use x as the input variable. Since h is what is varying in the slope formula, replace h by X and enter the slope formula in Y2.

Turn Y1 off since we are looking only at the output from Y2.

Press [2nd] [WINDOW] (TblSet) and choose the settings on the right.

(Since we are using the ASK feature, the settings for TblMin and ∆Tbl do not matter.)

Access the table with [2nd] [GRAPH] (TABLE) and either delete or type over any previous entries in the X column.

Let X (really h) take on values that move the nearby point on the left closer and closer to 8.

• Notice that after a certain point, the calculator switches your input values to scientific notation and displays rounded output values so that the numbers can fit on the screen in the space allotted for outputs of the table. You should position the cursor over each output value and record on paper as many decimal places as necessary in order to determine the limit from the left to the desired degree of accuracy.

Repeat the process, letting X (really h) take on values that move the nearby point on the right closer and closer to 8.

View the entire decimal value for each output and determine the limit from the right.

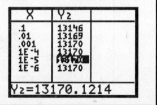

4.1.3 VISUALIZING THE LIMITING PROCESS (optional)   Program SECTAN can be

used to view secant lines between a point $(a, f(a))$ and some close points on a curve $y = f(x)$ and the tangent line at the point $(a, f(a))$. Using this program either before or after numerically finding the limit of the slopes can help you understand the numerical process.

We use the function giving the number of polio cases in 1949: $y = \dfrac{42183.911}{1 + 21484.253e^{-1.248911t}}$

where $t = 1$ on January 31, 1949, $t = 2$ on February 28, 1949, and so forth.

(Program SECTAN is given in the TI-82/83 Appendix and should be in your calculator before you work through the following illustration.)

Before using program SECTAN, the function must be entered in the Y1 location of the Y= list, and you *must* draw a graph of the function.
 	
Enter the function, using x as the input variable, in Y1.
	`Y1=42183.911/(1+`
	`21484.253e^(-1.2`
	`48911X))`
	`Y2=`
	`Y3=`
	`Y4=`
	`Y5=`
	`Y6=`

Since $t = 1$ represents January 31, 1949, 0 represents the beginning of 1949. The function gives the number of polio cases for the entire year, so we view the graph through December 31, 1949 ($t = 12$).

When you need to draw the graph of a function, you usually are given the input values in the statement of the problem in your text. *Always carefully read the problem before starting the solution process.*

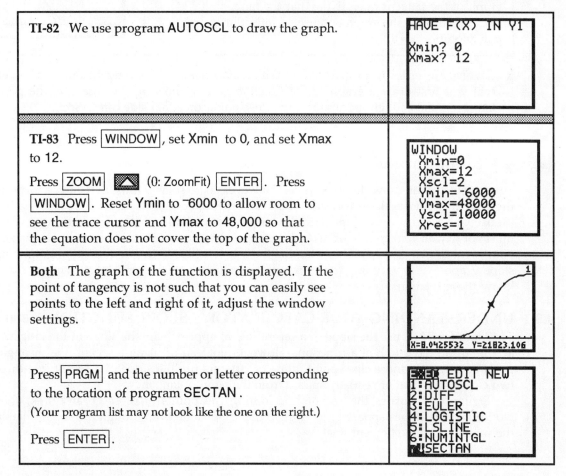

TI-82   We use program AUTOSCL to draw the graph.

```
HAVE F(X) IN Y1

Xmin? 0
Xmax? 12
```

TI-83   Press WINDOW, set Xmin to 0, and set Xmax to 12.

Press ZOOM ▲ (0: ZoomFit) ENTER. Press WINDOW. Reset Ymin to ⁻6000 to allow room to see the trace cursor and Ymax to 48,000 so that the equation does not cover the top of the graph.

```
WINDOW
  Xmin=0
  Xmax=12
  Xscl=2
  Ymin=-6000
  Ymax=48000
  Yscl=10000
  Xres=1
```

Both   The graph of the function is displayed. If the point of tangency is not such that you can easily see points to the left and right of it, adjust the window settings.

`X=8.0425532  Y=21823.106`

Press PRGM and the number or letter corresponding to the location of program SECTAN.

(Your program list may not look like the one on the right.)

Press ENTER.

```
EXEC EDIT NEW
1:AUTOSCL
2:DIFF
3:EULER
4:LOGISTIC
5:LSLINE
6:NUMINTGL
7:SECTAN
```

If you start this program and have forgotten to enter the function in Y1 or to draw a graph of the function, enter 2. Otherwise, type 1 and press ENTER.
 	
At the prompt, enter the input value of the tangent point. (For this example, $x = 8$.)
	`HAVE F(X)IN Y1...`
	`DRAW GRAPH OF F`
	
	`CONTINUE?`
	`YES(1) NO(2)1`
	`X-VALUE OF POINT`
	`OF TANGENCY? 8`

The next message that appears tells you to press enter to see secant lines drawn between the point of tangency and close points to the *left*. Press ENTER. (Five secant lines will draw ; they may take a little time to draw.)	
When you finish looking at the graph of the secant lines, press ENTER to continue. The next message that appears tells you to press enter to see secant lines drawn between the point of tangency and close points to the *right*. Press ENTER.	
Press ENTER to continue the program. You are next instructed to press ENTER to see a graph of the tangent line. (Note that the tangent line cuts through the graph because an inflection point occurs at $x \approx 8$.)	

- *Caution:* In order to properly view the secant lines and the tangent line, it is essential that you first draw a graph of the function clearly showing the function, the point of tangency, and enough space so that the close points on either side can be seen.

4.3 Slope Formulas

Your calculator can draw slope formulas. However, to do so, you must first enter a formula for the function whose slope formula you want the calculator to draw. Because you will probably be asked to draw slope formulas for functions whose equations you are not given, you must not rely on your calculator to do this for you. You should instead use technology to check your hand-drawn graphs and to examine the relationships between a function graph and its slope graph. It is very important in both this chapter and several later chapters that you know these relationships.

4.3.1 UNDERSTANDING YOUR CALCULATOR'S SLOPE FUNCTION   Both the TI-82 and the TI-83 use the slope of a secant line to approximate the slope of the tangent line at a point on the graph of a function. However, instead of using a secant line through the point of tangency and a close point, these calculators use the slope of a secant line through two close points that are equally spaced from the point of tangency.

Figure 7 illustrates the secant line joining the points $(a-k, f(a-k))$ and $(a+k, f(a+k))$. Notice that the slopes appear to be close to the same value even though the secant line is not the same line as the tangent line.

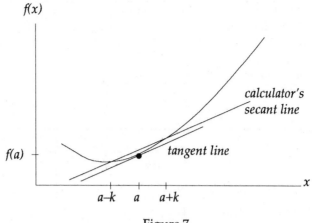

Figure 7

As k gets closer and closer to 0, the two points move closer and closer to a. Provided the slope of the tangent line exists, the limiting position of the secant line will be the tangent line.

The calculator's notation for the slope of the secant line shown in Figure 7 is

nDeriv(function, symbol for input variable, a, k)

Specifying the value of k is optional. If it is not given, the calculator automatically uses $k = 0.001$. Any smooth, continuous function will do, so let's investigate these ideas with the function $f(x) = x^3 - 4x^2 + 3.27x - 8.65$.

Enter $f(x) = x^3 - 4x^2 + 3.27x - 8.65$ in one of the locations of the Y= list, say Y1. Return to the home screen with [2nd] [MODE] (QUIT).	`Y1◼X^3-4X²+3.27X` `-8.65` `Y2=` `Y3=` `Y4=` `Y5=` `Y6=` `Y7=`
Suppose you want to find the slope of the secant line between the points $(^-1, f(^-1))$ and $(1, f(1))$. That is, you are finding the slope of the secant line between $(a-k, f(a-k))$ and $(a+k, f(a+k))$ for $a = 0$ and $k = 1$. Type the expression on the right. Access nDeriv(with [MATH] [8] (nDeriv(). Press [ENTER] and see that the slope of the secant line is 4.27.	`nDeriv(Y1,X,0,1)` `                4.27`
Press [2nd] [ENTER] (ENTRY), and edit the expression so that k changes from 1 to 0.1. Press [ENTER]. Again press [2nd] [ENTER] (ENTRY), and edit the expression so that k changes from 0.1 to 0.01. Press [ENTER]. Repeat the process for $k = 0.001$ and $k = 0.0001$. Did you get the values 3.28, 3.2701, 3.27001, and 3.27000001?	`nDeriv(Y1,X,0,1)` `                4.27` `nDeriv(Y1,X,0,.0` `1)` `             3.2701` `nDeriv(Y1,X,0,.0` `01)`

In the table on the next page, the first row lists some values of a, the input of the point of tangency, and the second row gives the slope of the tangent line at those values. (You will later learn how to find these exact values of the slope of the tangent line to $f(x)$ at various input values.)

Use your calculator to verify the values in the third through sixth rows that give the slope of the secant line between the points *(a-k, f(a-k))* and *(a+k, f(a+k))* for the indicated values of *k*. Find each secant line slope by calculating the value of nDeriv(Y1, X, *a*, *k*).

a = input of point of tangency	-1	2.3	12.82	62.7
slope of tangent line	14.27	0.74	393.7672	11295.54
slope of secant line, *k* = 0.1	14.28	0.75	393.7772	11295.55
slope of secant line, *k* = 0.01	14.2701	0.7401	393.7673	11295.5401
slope of secant line, *k* = 0.001	14.270001	0.740001	393.767201	11295.54
slope of secant line, *k* = 0.0001	14.27000001	0.74000001	393.7672	11295.54

You can see that the slope of the secant line is very close to the slope of the tangent line for small values of *k*. The slope of this secant line does a good job of approximating the slope of the tangent line when *k* is very small.

Now repeat the process, but do not include *k* in the instruction. That is, find the secant line slope by calculating nDeriv(Y1, X, *a*). Did you obtain the following?

slope of secant line   14.270001   0.740001   393.767201   11295.54

These values are those in the fifth row of the above table -- the values for *k* = 0.001. From this point forward, we use *k* = 0.001 and therefore do not specify *k* when evaluating nDeriv using the calculator.

Will the slope of this secant line always do a good job of approximating the slope of the tangent line when *k* is very small? Yes, it does, as long as the instantaneous rate of change exists at the input value (*a*) at which you evaluate nDeriv. When the instantaneous rate of change does not exist at a point, nDeriv should *not* be used to approximate something that does not have a value!

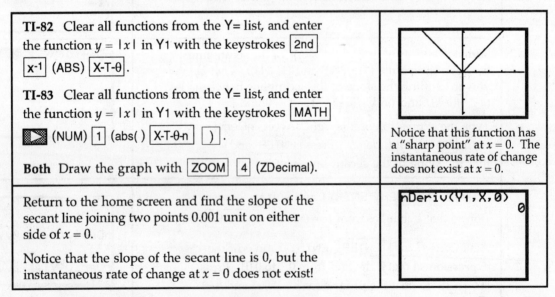

TI-82 Clear all functions from the Y= list, and enter the function $y = |x|$ in Y1 with the keystrokes [2nd] [x⁻¹] (ABS) [X-T-θ].

TI-83 Clear all functions from the Y= list, and enter the function $y = |x|$ in Y1 with the keystrokes [MATH] [▶] (NUM) [1] (abs( ) [X-T-θ-n] [ )].

Both Draw the graph with [ZOOM] [4] (ZDecimal).

Notice that this function has a "sharp point" at $x = 0$. The instantaneous rate of change does not exist at $x = 0$.

Return to the home screen and find the slope of the secant line joining two points 0.001 unit on either side of $x = 0$.

Notice that the slope of the secant line is 0, but the instantaneous rate of change at $x = 0$ does not exist!

- Be certain the instantaneous rate of change exists at a point before using nDeriv. Two places where nDeriv usually does *not* give correct results *for the instantaneous rate of change* are at sharp points and the joining point(s) of piecewise continuous functions.

- Provided the instantaneous rate of change exists at a point, we use the secant line slope nDeriv to provide a good approximation to the slope of the tangent line at that point. Since the slope of the tangent line is the slope of the curve which is the derivative of the function, we call nDeriv the calculator's *numerical derivative*.

4.3.2 DERIVATIVE NOTATION AND CALCULATOR NOTATION

You can often see a pattern in a table of values for the slopes of a function at indicated values of the input variable and discover a formula for the slope (derivative). The process of calculating the slopes uses the calculator's numerical derivative, nDeriv(f(X), X, X). The correspondence between the derivative notation $\frac{d\,f(x)}{dx}$ and the calculator's notation nDeriv(f(X), X, X) is

nDeriv ($f(x)$, ☒ , △)

$\frac{d}{d\boxed{x}}$ $f(\triangle x)$)

○ indicates we are taking a derivative or slope.

□ indicates the letter corresponding to the name of the input variable.

△ indicates the value of the input variable at which the slope is calcualated.

Suppose you are asked to construct a table of values of $f'(x)$ where $f(x) = x^2$ evaluated at different values of x. Two methods of doing this are illustrated below:

Return to the home screen and type the expression on the right. Access nDeriv(with MATH 8 (nDeriv(). nDeriv(X², X, 2) $\approx \frac{dy}{dx}$ for $y = x^2$ evaluated at $x = 2$.	`nDeriv(X²,X,2)` `                4`

Recall the last entry, and edit the expression nDeriv(X², X, 2) by changing the 2 to a ⁻3.
 	
Press ENTER. Continue with this process until you have found all the slope values.
	`nDeriv(X²,X,⁻3)`
	`            ⁻6`
	`nDeriv(X²,X,⁻2)`
	`            ⁻4`
	`nDeriv(Y₁,X,⁻1)`
	`            ⁻1`
	`nDeriv(X²,X,0)`

You might prefer to use the TABLE. If so, recall that you must have the expression being evaluated in the Y= list.
 	
If you enter the slope formula as indicated on the right, you will only have to change Y1 when you work with a different function.
	`Y₁▣X²`
	`Y₂▣nDeriv(Y₁,X,X`
	`)`
	`Y₃=`
	`Y₄=`
	`Y₅=`
	`Y₆=`
	`Y₇=`

You can either type in the x-values using the ASK feature of the table or you can set TblMin = ⁻3 and ΔTbl = 1 with the AUTO setting chosen.	`TABLE SETUP`
	`  TblMin=⁻3`
	`  ΔTbl=1`
	`Indpnt: AUTO Ask`
	`Depend: AUTO Ask`

Use the table and evaluate the function and the numerical derivative at $x = $ ⁻3, ⁻2, ⁻1, 0, 1, 2, and 3.
 	
Determine a relationship (pattern) between the slopes and the values of x.
 	
Function values appear in the Y1 column and slopes appear in the Y2 column.
	X | Y₁ | Y₂ table:
	`⁻3  9   ⁻6`
	`⁻2  4   ⁻4`
	`⁻1  1   ⁻2`
	`0   0   0`
	`1   1   2`
	`2   4   4`
	`3   9   6`
	`X=⁻3`

- If you have difficulty determining a pattern, enter the x-values at which you are evaluating the slope in list L1 and the values of nDeriv in list L2. Draw a scatter plot of the x-values and the slope values. The shape of the scatter plot should give you a clue as to the equation of the slope formula. If not, try drawing another scatter plot where L1 contains the values of $y = f(x)$ and L2 contains the calculated slope formula values. Note that this method might help only if you consider a variety of values for x in list L1.

- The TI-82 and the TI-83 only calculate approximate numerical values of slopes -- they do not give the slope in formula form.

4.3.3 DRAWING TANGENT LINES FROM THE GRAPHICS SCREEN Chapter 3 of this *Guide* (specifically, Section 3.3.2) presented a method of drawing tangent lines from the home screen. We now examine another method for drawing tangent lines, this time using the graphics screen. You may not find this method as useful as the previous one, however, because the point at which the tangent line is drawn depends on the horizontal settings in the window.

We illustrate this method of drawing tangent lines with $f(x) = 2\sqrt{x - 5}$. Without the context of a real-world situation, how do you know what input values to consider? The answer is that you need to call upon your knowledge of functions. Remember that we graph only real numbers. If the quantity under the square root symbol is negative, the output of $f(x)$ is not a real number. We therefore know that x must be greater than or equal to 5. Many different horizontal views will do, but we choose to use $0 \le x \le 15$. You can use previously-discussed methods to set height of the window, or you can use the one given below.

Enter $f(x) = 2\sqrt{x - 5}$ in Y1. The parentheses around the $x - 5$ are necessary to include the entire quantity under the square root.	Y1■2√(X-5) Y2= Y3= Y4= Y5= Y6= Y7= Y8=
Set the window to that shown on the right. Press GRAPH.	WINDOW FORMAT Xmin=0 Xmax=15 Xscl=5 Ymin=-5 Ymax=10 Yscl=1
TI-82 With the graph on the screen, press 2nd PRGM (DRAW) 5 (Tangent( ). Use ▶ or ◀ to move to some point on the curve. Press ENTER.	 X=7.9787234 Y=3.4517957
The tangent line is drawn at the position of the cursor. At the bottom of the screen, the TI-82 also gives the slope of the tangent line at that point.	 dy/dx=.57940857

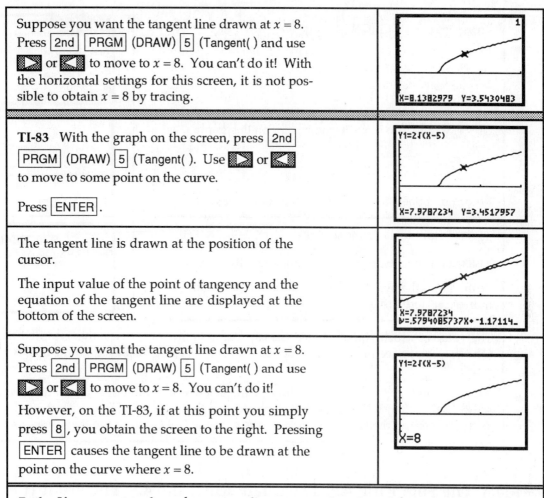

Suppose you want the tangent line drawn at $x = 8$. Press ⟨2nd⟩ ⟨PRGM⟩ (DRAW) ⟨5⟩ (Tangent( ) and use ⟨▶⟩ or ⟨◀⟩ to move to $x = 8$. You can't do it! With the horizontal settings for this screen, it is not possible to obtain $x = 8$ by tracing.

TI-83 With the graph on the screen, press ⟨2nd⟩ ⟨PRGM⟩ (DRAW) ⟨5⟩ (Tangent( ). Use ⟨▶⟩ or ⟨◀⟩ to move to some point on the curve.

Press ⟨ENTER⟩.

The tangent line is drawn at the position of the cursor.

The input value of the point of tangency and the equation of the tangent line are displayed at the bottom of the screen.

Suppose you want the tangent line drawn at $x = 8$. Press ⟨2nd⟩ ⟨PRGM⟩ (DRAW) ⟨5⟩ (Tangent( ) and use ⟨▶⟩ or ⟨◀⟩ to move to $x = 8$. You can't do it!

However, on the TI-83, if at this point you simply press ⟨8⟩, you obtain the screen to the right. Pressing ⟨ENTER⟩ causes the tangent line to be drawn at the point on the curve where $x = 8$.

Both If you want to draw the tangent line at a certain value of the input variable that is not a possible trace value, return to the home screen and use the method given in Section 3.3.2 of this *Guide*. (On the TI-83, either method can be used.) It is very important to remember the situations discussed in that section in which the instantaneous rate of change does not exist, but yet the calculator's tangent line draws on the screen.

4.3.4 CALCULATING $\dfrac{dy}{dx}$ AT SPECIFIC INPUT VALUES   Section 4.3.1 of this *Guide*

examined the calculator's numerical derivative nDeriv(Y1, X, *a*) and illustrated that it gives a good approximation of the slope of the tangent line at points where the instantaneous rate of change exists. You can also evaluate the calculator's numerical derivative from the graphics screen using the **CALC** menu. However, instead of being called nDeriv in that menu, it is called $\dfrac{dy}{dx}$. We illustrate its use with the function $f(x) = 2\sqrt{x - 5}$.

Enter $f(x) = 2\sqrt{x - 5}$ in Y1, and draw a graph of $f(x)$.

(Refer to Section 4.3.3 of this *Guide*. If you have the graph on the screen with the tangent line from the previous section, retype the 2 in $f(x)$ and the graph will draw as on the right.)

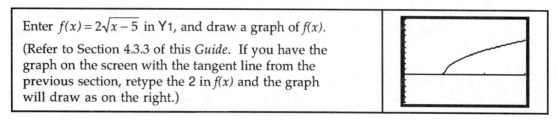

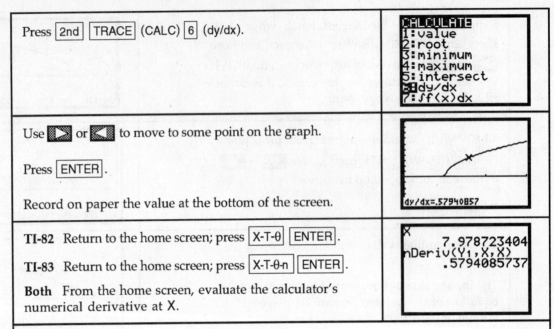

Press ⬛2nd⬛ ⬛TRACE⬛ (CALC) ⬛6⬛ (dy/dx).	**CALCULATE** 1:value 2:root 3:minimum 4:maximum 5:intersect 6:dy/dx 7:∫f(x)dx
Use ⬛▶⬛ or ⬛◀⬛ to move to some point on the graph. Press ⬛ENTER⬛. Record on paper the value at the bottom of the screen.	dy/dx=.57940857

TI-82 Return to the home screen; press ⬛X-T-θ⬛ ⬛ENTER⬛.
 	
TI-83 Return to the home screen; press ⬛X-T-θ-n⬛ ⬛ENTER⬛.
 	
Both From the home screen, evaluate the calculator's numerical derivative at X.
	X
	     7.978723404
	nDeriv(Y₁,X,X)
	     .5794085737

The value at which you evaluated the calculator's numerical derivative is stored in X. (The values you see probably will not be the same as those displayed on the above screens.)

As you would expect, the two values of the calculator's numerical derivative are the same. Other than fewer digits being printed on the graphics screen with the value of *dy/dx*, nDeriv on the home screen and dy/dx on the CALC screen are the same.

📖 4.4 The Sum Rule, 4.5 The Chain Rule, and 4.6 The Product Rule

If you have time, it is always a good idea to check your answer. Although your calculator cannot give you a general rule for the derivative of a function, you can use graphical and numerical techniques to check your derivative formula answers. These same procedures apply when you check your results after applying the Sum Rule, the Chain Rule, or the Product Rule.

4.4.1 NUMERICALLY CHECKING SLOPE FORMULAS   When you use a formula to find the derivative of a function, it is possible to check your answer using the calculator's numerical derivative nDeriv. The basic idea of the checking process is that if you evaluate your derivative and the calculator's numerical derivative at several randomly chosen values of the input variable and the output values are very close to the same values, your derivative is *probably* correct.

The average yearly fuel consumption per car in the United States from 1980 through 1990 can be modeled by $g(t) = 0.775t^2 - 140.460t + 6868.818$ gallons per car where t is the number of years since 1900. Applying the sum, power, and constant multiplier rules for derivatives, suppose you determine $g'(t) = 1.55t - 140.460$ gallons per year per car. We now numerically check this answer.

(As we have mentioned several times, if you have found a model from data, you should have the complete model, not the rounded one given by $g(t)$, in the Y= list of the calculator.)

Enter the function you are taking the derivative of in Y1, the calculator's derivative in Y2, and your derivative formula, $\frac{dg}{dt} = g'(t)$, in Y3.	`Y1■.775058275058`
	`36X^2+-140.45990`
	`675992X+6868.818`
	`1818188`
	`Y2■nDeriv(Y1,X,X`
	`)`
	`Y3■1.55X-140.46`
	`Y4=`

Since the $g(t)$ model represents average fuel consumption per car where $t = 80$ in 1980, it makes sense to check using only whole number values of t greater than or equal to 80 (that are now denoted by x since we are using the Y= list).	In **TABLE SETUP**, choose **ASK** in the independent variable location. Turn off Y1 since you are checking to see if Y2 ≈ Y3.

Press 2nd GRAPH (TABLE) and check to see that Y2 and Y3 are very close to the same values for at least four values of x.
 	
(Recall that we are using a rounded derivative in Y3. This accounts for most of the differences in the two columns of output values.)
	`X    Y2      Y3`
	`80   -16.45  -16.46`
	`81   -14.9   -14.91`
	`82   -13.35  -13.36`
	`83   -11.8   -11.81`
	`84   -10.25  -10.26`
	`85   -8.7    -8.71`
	`X=`

If the two columns of output values are *not* very close to the same, you have either incorrectly entered a function in the Y= list or you have made a mistake in your derivative formula.

4.4.2 GRAPHICALLY CHECKING SLOPE FORMULAS   Another method of checking your answer for a slope formula (derivative) is to draw the graph of the calculator's numerical derivative and draw the graph of your derivative. If the graphs appear identical *in the same viewing window*, your derivative is probably correct.

We again use the fuel consumption functions from Section 4.4.1 of this *Guide*.

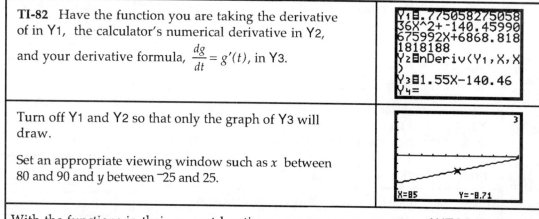

TI-82 Have the function you are taking the derivative of in Y1, the calculator's numerical derivative in Y2, and your derivative formula, $\frac{dg}{dt} = g'(t)$, in Y3.	`Y1■.775058275058`
	`36X^2+-140.45990`
	`675992X+6868.818`
	`1818188`
	`Y2■nDeriv(Y1,X,X`
	`)`
	`Y3■1.55X-140.46`
	`Y4=`

Turn off Y1 and Y2 so that only the graph of Y3 will draw. Set an appropriate viewing window such as x between 80 and 90 and y between ⁻25 and 25.	`X=85          Y=-8.71`

With the functions in their current locations, you cannot use program AUTOSCL since it only draws the graph of the function in Y1. You could change the location of the functions and use the program, or you could use ZOOM 4 (ZDecimal). You probably won't see much, but press TRACE and ▼ until you see a "3" in the upper-right-hand corner of the screen. Press ▶ and ◀ several times and watch the y-coordinates. Reset Ymin and Ymax in the window based on the y-value observations, and press GRAPH. If necessary, repeat this process until you see a good graph of Y3.

Now, turn off Y3, turn on Y2, and draw the graph of Y2 *in the same viewing window.*

(The graph of the calculator's numerical derivative takes slightly longer to draw because your calculator computes the output before plotting each point.)

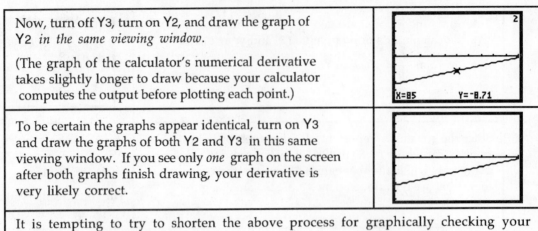

To be certain the graphs appear identical, turn on Y3 and draw the graphs of both Y2 and Y3 in this same viewing window. If you see only *one* graph on the screen after both graphs finish drawing, your derivative is very likely correct.

It is tempting to try to shorten the above process for graphically checking your derivative formula by drawing only the graphs of your derivative and the calculator's derivative at the same time without first graphing each separately. If your derivative is such that it cannot be seen in the viewing window in which you see the calculator's derivative or vice versa, you will see only one graph and think that your slope formula is correct. It is better to perform a numerical check on the derivatives than to incorrectly use the graphical checking process.

TI-83 Have the function you are taking the derivative of in Y1, the calculator's numerical derivative in Y2, and your derivative formula, $\frac{dg}{dt} = g'(t)$, in Y3.

Turn off Y1 so that only the graphs of Y2 and Y3 will draw. Use ▼ until you reach the Y3 location, and press ◀ ◀ to move the cursor over the "\" mark to the left of Y3. Press ENTER and notice the blinking cursor changes to a heavier slanted line.

Press WINDOW, set Xmin to 80, and set Xmax to 90. Press ZOOM ▲ (0: ZoomFit) ENTER.

Carefully watch the screen. If you see the first graph (Y2) draw and then see the darker graph (Y3) draw *on top of* the first graph, your derivative is probably correct. You must see both graphs draw for this to method to work as a check of your answer!

If you want to see the graphs draw again, just change a value in the window and press GRAPH.

It is not necessary to reset the darker line to the left of Y3 for future graphing. When you clear the equation in Y3, the setting returns to the normal slanted line.

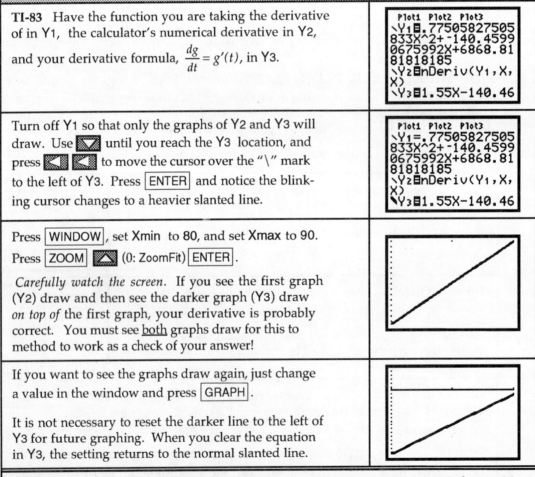

Both When trying to determine an appropriate viewing window, read the problem again; it will likely indicate the values for Xmin and Xmax. Also use your knowledge of the general shape of the function being graphed to know what you should see.

Chapter 5   Analyzing Change: Extrema and Points of Inflection

📖 5.1 Optimization

Your calculator can be very helpful in checking your analytic work when you find optimal points and points of inflection. When you are not required to show work using derivatives or when a very good approximation to the exact answer is all that is required, it is a very simple process to use your calculator to find optimal points and inflection points.

5.1.1 FINDING X-INTERCEPTS OF SLOPE GRAPHS   Where the graph of a function has a local maximum or minimum, the slope graph has a horizontal tangent. Where the tangent line is horizontal, the derivative of the function is zero. Thus, finding where the slope graph *crosses* the input axis is the same as finding where a relative maximum or a relative minimum occurs.

Consider, for example, the model for a cable company's revenue for the 26 weeks after it began a sales campaign:

$$R(x) = {}^{-}3x^4 + 160x^3 - 3000x^2 + 24{,}000x \text{ dollars}$$

where x is the number of weeks since the cable company began sales.

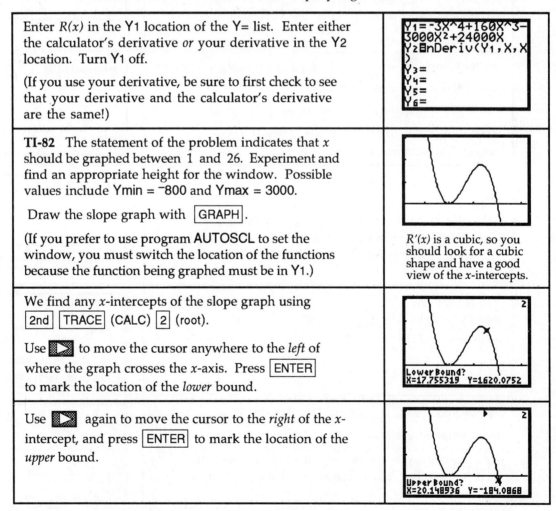

Enter $R(x)$ in the Y1 location of the Y= list. Enter either the calculator's derivative *or* your derivative in the Y2 location. Turn Y1 off. (If you use your derivative, be sure to first check to see that your derivative and the calculator's derivative are the same!)	```Y1=-3X^4+160X^3-3000X²+24000X Y2=nDeriv(Y1,X,X) Y3= Y4= Y5= Y6=```
TI-82 The statement of the problem indicates that x should be graphed between 1 and 26. Experiment and find an appropriate height for the window. Possible values include Ymin = ⁻800 and Ymax = 3000. Draw the slope graph with GRAPH . (If you prefer to use program **AUTOSCL** to set the window, you must switch the location of the functions because the function being graphed must be in Y1.)	$R'(x)$ is a cubic, so you should look for a cubic shape and have a good view of the x-intercepts.
We find any x-intercepts of the slope graph using 2nd TRACE (CALC) 2 (root). Use ▶ to move the cursor anywhere to the *left* of where the graph crosses the x-axis. Press ENTER to mark the location of the *lower* bound.	Lower Bound? X=17.755319 Y=1620.0752
Use ▶ again to move the cursor to the *right* of the x-intercept, and press ENTER to mark the location of the *upper* bound.	Upper Bound? X=20.148936 Y=-184.0868

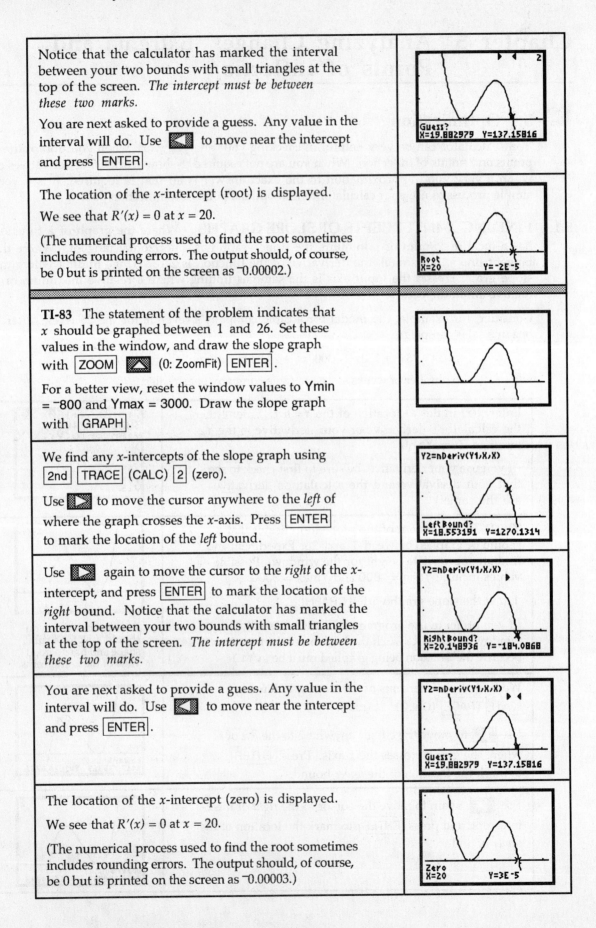

Notice that the calculator has marked the interval between your two bounds with small triangles at the top of the screen. *The intercept must be between these two marks.*

You are next asked to provide a guess. Any value in the interval will do. Use ◀ to move near the intercept and press ENTER .

The location of the *x*-intercept (root) is displayed.

We see that $R'(x) = 0$ at $x = 20$.

(The numerical process used to find the root sometimes includes rounding errors. The output should, of course, be 0 but is printed on the screen as ⁻0.00002.)

TI-83 The statement of the problem indicates that *x* should be graphed between 1 and 26. Set these values in the window, and draw the slope graph with ZOOM ▲ (0: ZoomFit) ENTER .

For a better view, reset the window values to Ymin = ⁻800 and Ymax = 3000. Draw the slope graph with GRAPH .

We find any *x*-intercepts of the slope graph using 2nd TRACE (CALC) 2 (zero).

Use ▶ to move the cursor anywhere to the *left* of where the graph crosses the *x*-axis. Press ENTER to mark the location of the *left* bound.

Use ▶ again to move the cursor to the *right* of the *x*-intercept, and press ENTER to mark the location of the *right* bound. Notice that the calculator has marked the interval between your two bounds with small triangles at the top of the screen. *The intercept must be between these two marks.*

You are next asked to provide a guess. Any value in the interval will do. Use ◀ to move near the intercept and press ENTER .

The location of the *x*-intercept (zero) is displayed.

We see that $R'(x) = 0$ at $x = 20$.

(The numerical process used to find the root sometimes includes rounding errors. The output should, of course, be 0 but is printed on the screen as ⁻0.00003.)

Both Now, you must determine if the derivative graph crosses or just touches the *x*-axis at the location to the left of this intercept.

Use **Zbox** as many times as necessary to magnify that portion of the graph to see what happens there. (See Section 3.3.1 of this *Guide*.)

```
X=13.5          Y=-248.3871
```

After using **Zbox** several times, we see that the graph just touches and does not cross the *x*-axis near *x* = 10.

Using the information gained from $R'(x)$ and the graph of $R(x)$, we find that revenue was greatest 20 weeks after the cable company began sales.

```
X=9.9992589  Y=.18629785
```

5.1.2 FINDING OPTIMAL POINTS   Once you draw a graph of a function that clearly shows any optimal points, finding the location of those high points and low points is an easy task for your calculator. When a relative maximum or a relative minimum exists at a point, your calculator can find it in a few simple steps. We again use the cable company revenue equation, $R(x)$, from Section 5.1.1.

TI-82   Enter $R(x)$ in the Y1 location of the Y= list.

(Any location will do, but if you have other equations in the graphing list, turn them off or clear them.)

Next, draw a graph of the revenue, $R(x)$.

```
Y1 = -3X^4+160X^3-
3000X²+24000X
Y2=nDeriv(Y1,X,X
)
Y3=
Y4=
Y5=
Y6=
```

The input *x* should be graphed between 1 and 26.

You could experiment and find an appropriate height for the window, or you could use program **AUTOSCL**. (Remember, if you use the program, the function being graphed must be in Y1.)

Prepare to find the local (relative) maximum by pressing [2nd] [TRACE] (CALC) [4] (maximum).

Next, press [▶] to move the blinking cursor that appears to a position to the *left* of the high point. Press [ENTER] to mark the *lower* bound for *x*.

```
Lower Bound?
X=16.159574  Y=75029.446
```

Next, press [▶] to move the blinking cursor to a position to the *right* of the high point. Press [ENTER] to mark the *upper* bound for *x*.

```
Upper Bound?
X=22.808511  Y=73308.489
```

Use 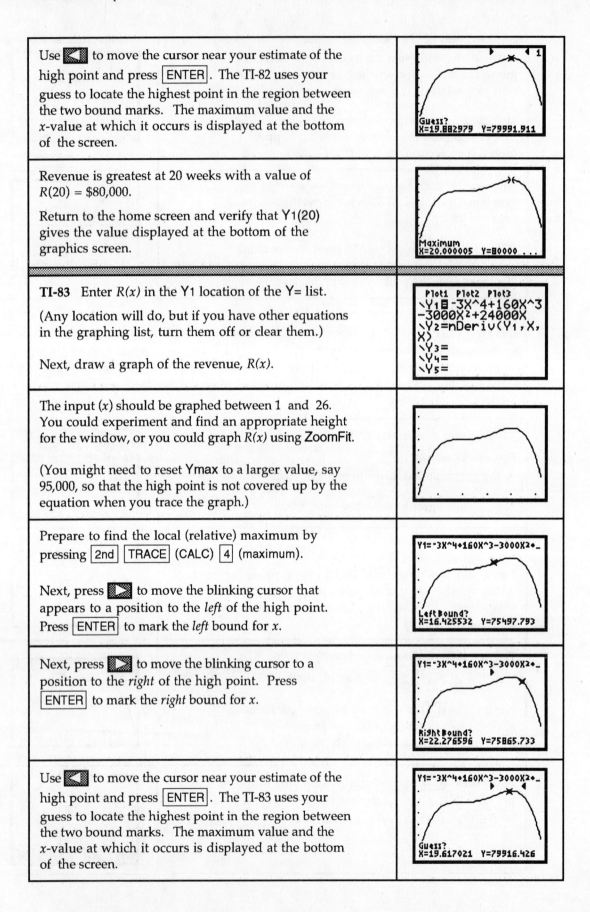 to move the cursor near your estimate of the high point and press ENTER . The TI-82 uses your guess to locate the highest point in the region between the two bound marks.   The maximum value and the x-value at which it occurs is displayed at the bottom of the screen.

Guess?
X=19.882979 Y=79991.911

Revenue is greatest at 20 weeks with a value of $R(20) = \$80,000$.

Return to the home screen and verify that Y1(20) gives the value displayed at the bottom of the graphics screen.

Maximum
X=20.000005 Y=80000 ...

TI-83   Enter $R(x)$ in the Y1 location of the Y= list.

(Any location will do, but if you have other equations in the graphing list, turn them off or clear them.)

Next, draw a graph of the revenue, $R(x)$.

Plot1 Plot2 Plot3
\Y1◻-3X^4+160X^3
-3000X²+24000X
\Y2=nDeriv(Y1,X,
X)
\Y3=
\Y4=
\Y5=

The input (x) should be graphed between 1 and 26. You could experiment and find an appropriate height for the window, or you could graph $R(x)$ using ZoomFit.

(You might need to reset Ymax to a larger value, say 95,000, so that the high point is not covered up by the equation when you trace the graph.)

Prepare to find the local (relative) maximum by pressing 2nd TRACE (CALC) 4 (maximum).

Next, press 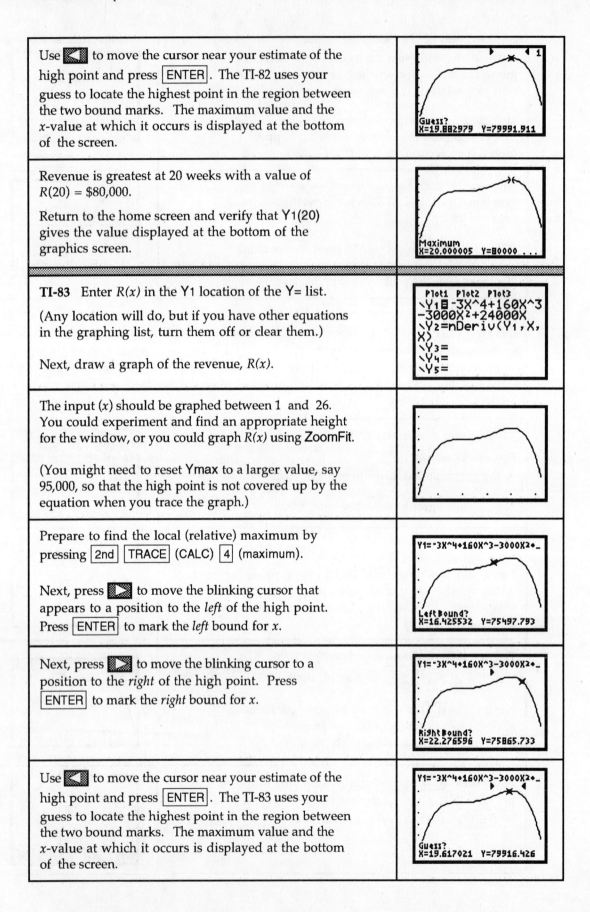 to move the blinking cursor that appears to a position to the *left* of the high point. Press ENTER to mark the *left* bound for x.

Y1=-3X^4+160X^3-3000X²◆
Left Bound?
X=16.425532 Y=75497.793

Next, press 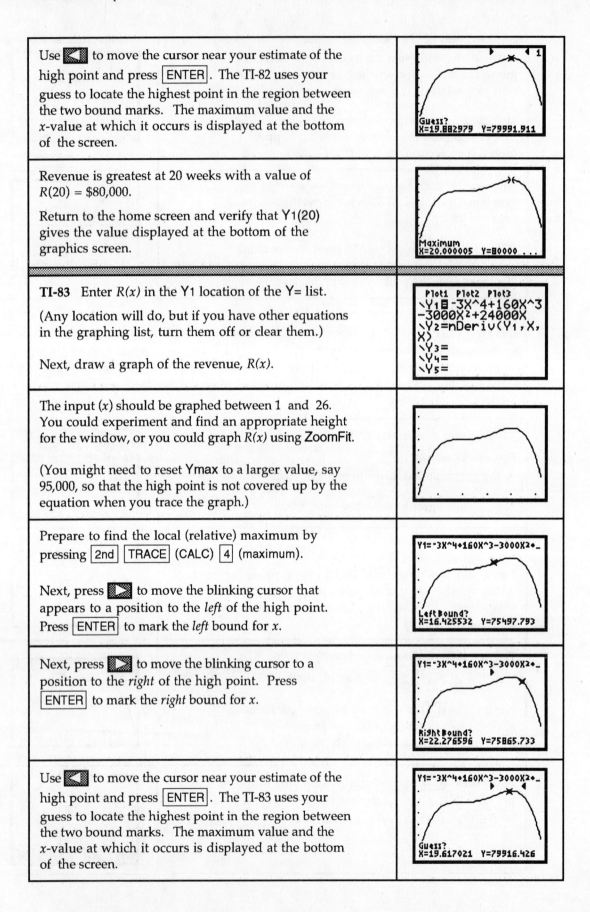 to move the blinking cursor to a position to the *right* of the high point. Press ENTER to mark the *right* bound for x.

Y1=-3X^4+160X^3-3000X²◆
Right Bound?
X=22.276596 Y=75865.733

Use 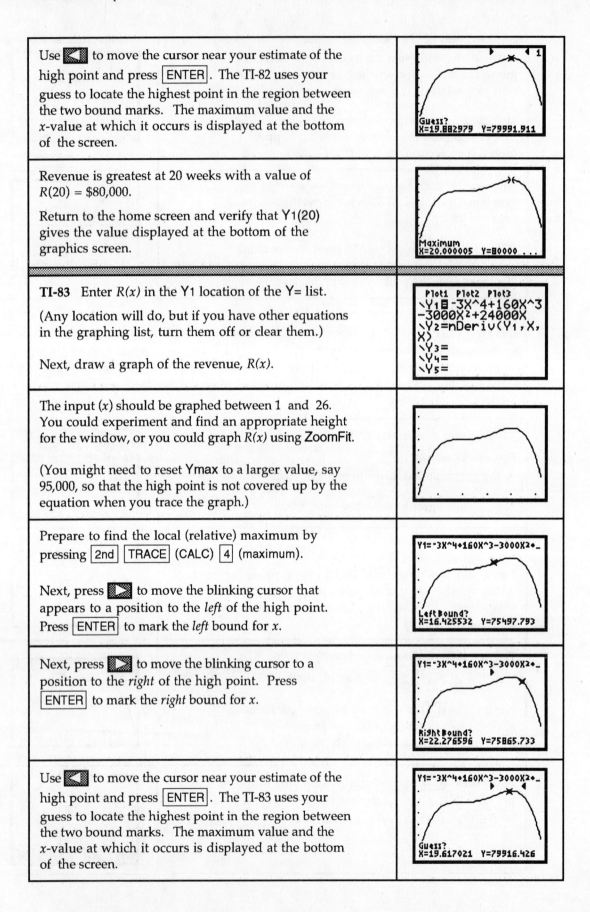 to move the cursor near your estimate of the high point and press ENTER . The TI-83 uses your guess to locate the highest point in the region between the two bound marks. The maximum value and the x-value at which it occurs is displayed at the bottom of the screen.

Y1=-3X^4+160X^3-3000X²◆
Guess?
X=19.617021 Y=79916.426

Revenue is greatest at 20 weeks with a value of $R(20) = \$80,000$. Return to the home screen and verify that Y1(20) gives the value displayed at the bottom of the graphics screen.	Maximum X=20.000001 Y=80000
Both What about the other part of the curve that may contain a peak? Follow the same procedure as indicated above to investigate the curve on either side of $x = 10$.	Guess? X=10.574468 Y=70007.257
Notice that the calculator returns the maximum at the right end of the interval. This indicates that there is not a relative maximum value in the interval we marked to test.	Maximum X=14.297864 Y=72151.939

- The methods of this section also apply to finding relative or local minimum values of a function. The only difference is that to find the minimum instead of the maximum, initially press ⟨2nd⟩ ⟨TRACE⟩ (CALC) ⟨3⟩ (minimum).

📖 5.2 Inflection Points

As was the case with optimal points, your calculator can be very helpful in checking your analytic work when you find points of inflection. You can also use the methods illustrated in Section 5.1.2 of this *Guide* to find the location of any maximum or minimum points on the graph of the first derivative to find the location of any inflection points for the function.

5.2.1 FINDING X-INTERCEPTS OF A SECOND DERIVATIVE GRAPH   We first look at using the analytic method of finding inflection points -- finding where the graph of the second derivative of a function *crosses* the input axis.

To illustrate, consider a model for the percentage of students graduating from high school in South Carolina from 1982 through 1990 who entered post-secondary institutions:

$$f(x) = {}^{-}0.1057x^3 + 1.355x^2 - 3.672x + 50.792 \text{ percent}$$

where $x = 0$ in 1982.

Enter $f(x)$ in the Y1 location of the Y= list, the first derivative in Y2 and the second derivative of the function in Y3. Turn off Y1 and Y2.	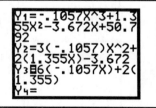

The problem says the model is for 1982 through 1990 which corresponds to $0 \le x \le 8$. Thus, you are told the horizontal view. Choose an appropriate vertical view -- possibly y between $^-4$ and 4. Graph $f''(x)$ Remember that you should be able to clearly see any optimal points. Leave room at the bottom of the screen so that trace coordinates will not block your view of any important points on the graph.	 Since the second derivative is a line and we need to find the x-intercept, this is a "good" graph.
Use the methods illustrated in 5.1.1 of this *Guide* to find where the second derivative graph crosses the x-axis. (The TI-82 calls this intercept a "root" while the TI-83 calls it a "zero". Both are acceptable terms.)	Root X=4.2731 Y=0

If you are asked to give the inflection *point* of $f(x)$, you should give both an x-value and a y-value.
 	
Return to the home screen and type X. The x-value you just found is stored in the x-location until you change it by tracing, using a menu item from the CALC menu, etc. Find the y-value by substituting this x-value in the function located in Y1.
	X
	       4.273099968
	Y1(X)
	       51.59548653

Next, look a graph of the function and verify that there does appear to be an inflection point at $x \approx 4.27$. To do this, turn off Y3 and turn on Y1. Draw the graph of Y1.	1 X=4.2553191 .Y=51.557826
In this problem, it is difficult to find a window that shows a good graph of both the function and its derivatives. However, if you draw a graph of all three, you can roughly see that the location of the inflection point of the function occurs at the location of the maximum of the first derivative and at the location of the x-intercept of the second derivative.	

Your calculator usually draws an accurate graph of the first derivative of a function when you use nDeriv. However, neither the TI-82 nor the TI-83 has a built-in method to calculate or graph $f''(x)$, the second derivative.

As illustrated below, you can try to use nDeriv($f'(x)$) for $f''(x)$. Be cautioned, however, that nDeriv($f'(x)$) sometimes "breaks down" and gives invalid results. If this should occur, the graph of nDeriv($f'(x)$) appears very jagged and this method should not be used.

Enter $f(x)$ in the Y1 location of the Y= list, your calculator's numerical derivative in the Y2 location, and the derivative of the calculator's first derivative in the Y3 location. Turn off Y1.	Y1=-.1057X^3+1.3 55X²-3.672X+50.7 92 Y2◻nDeriv(Y1,X,X) Y3◻nDeriv(Y2,X,X) Y4=

| Draw the graphs in a viewing window that gives a good view of the slope graph, $f'(x)$ in Y2, and its derivative, $f''(x)$ in Y3. The graph on the right was drawn with $0 \le x \le 8$ and $^-4 \le y \le 4$.

(Note that Y3 appears to give good results here.) | |
| Use the methods of 5.1.1 of this *Guide* to find the x-intercept of the second derivative. (It takes a little longer here than it did before!)

Be sure to use ▼ to move to the graph of the line before giving the lower (left) bound for the root (zero). | Root
X=4.2730999 Y=0 |

5.2.2 FINDING INFLECTION POINTS WITH YOUR CALCULATOR   Remember

that an inflection point is a point of greatest or least slope. Whenever finding the second derivative of a function is tedious or you do not need an exact answer from an analytic solution, you can very easily find an inflection point of a function by finding where the first derivative of the function has a maximum or minimum value.

We illustrate this process with the function giving the number of polio cases in 1949:

$$y = \frac{42183.911}{1 + 21484.253e^{-1.248911t}} \text{ where } t = 1 \text{ on January 31, 1949, } t = 2 \text{ on February 28, 1949, etc.}$$

Enter the function in the Y1 location of the Y= list and the calculator's numerical derivative in the Y2 location. Turn off Y1.	Y1=42183.911/(1+ 21484.253e^(-1.2 48911X)) Y2=nDeriv(Y1,X,X) Y3= Y4= Y5=
Graph Y2 in an appropriate viewing window with x between 0 and 12 and something like y between $^-3000$ and 15,000 with tick marks every 1000 units. Remember that you should be able to clearly see any optimal points. Leave room at the bottom of the screen so that trace coordinates will not block your view of any important points on the graph.	
Use the methods illustrated in 5.1.2 of this *Guide* to find the maximum of the slope graph. The x-value of the maximum of the slope graph is the x-value of the inflection point of the function.	Maximum X=7.9870255 Y=13170.986
If you are asked to give the inflection point, you should give both an x-value and a y-value. Find the y-value by substituting this x-value in the function located in Y1. The rate of change at that time is obtained by substituting this x-value in Y2.	X 7.987025459 Y1(X) 21092.04435 Y2(X) 13170.98591

Chapter 6 Accumulating Change: Limits of Sums and the Definite Integral

6.1 Results of Change

We have thus far seen how to use the calculator to work with rates of change. In this chapter we consider the results of change. Your calculator has many useful features that will assist you in your study of the accumulation of change.

6.1.1 APPROXIMATIONS WITH LEFT RECTANGLES   The calculator's lists can be used to perform the calculations needed to approximate, using left rectangles, the area between the horizontal axis, a (non-negative) rate of change function, and two input values.

Consider, for example, a model for the number of customers per minute who came to a Saturday sale at a large department store between 9 a.m. and 9 p.m.:

$$c(m) = (4.58904*10^{-8})\, m^3 - (7.78127*10^{-5})\, m^2 + 0.03303\, m + 0.88763$$

customers per minute where m is the number of minutes after 9 a.m.

Enter this model in the Y1 location of the Y= list. (You must use x as the input variable. If you have other equations in the graphing list, clear them.) (Remember that "10 to a power" is denoted by **E** on the calculator. Access **E** with [2nd] [,]. Use this symbol or the 10s with powers when entering Y1.)	Y1■(4.58904E-8)X ^3-(7.78127E-5)X ²+.03303X+.88763 Y₂= Y₃= Y₄= Y₅=
Suppose we want to estimate the total number of customers who came to the sale between $x = 0$ and $x = 660$ (12 hours) with 12 rectangles and $\Delta x = 60$. Enter these x-values in list L1. Recall that a quick way to do this is to have L1 darkened and then type [2nd] [STAT] (LIST) [5] (seq( ) [X,T,θ] [,] [X,T,θ] [,] 0 [,] 660 [,] 60 [ )] [ENTER]. (See Section 2.1.1 of this *Guide*.)	L1 L2 L3 360 420 480 540 600 ──────── L₁(12)=660 Notice that 9 p.m. is 720 minutes after 9 a.m. When using *left*-rectangle areas, the *rightmost* data point is not included.
Enter the $c(x)$ values calculated from the model in list L2 by pressing ▓▲ (to darken the name of this list) and typing Y1(L1) with the keystrokes: **TI-82** [2nd] [VARS] (Y-VARS) [1] (Function) [1] (Y1) [( ] [2nd] [1] (L1) [ )] [ENTER] **TI-83** [VARS] (Y-VARS) [▶] [1] (Function) [1] (Y1) [( ] [2nd] [1] (L1) [ )] [ENTER]	L1 L2 L3 0 .88763 60 2.5992 120 3.81 180 4.5795 240 4.9672 300 5.0325 360 4.835 L₂=Y₁(L₁) List L2 contains the *heights* of the 12 rectangles.

Both Since the *width* of each rectangle is 60, the area of each rectangle is 60*height*. Enter the *areas* of the 12 rectangles in list L3 by using [▲▲] to darken the name of list L3 and then typing 60L2 with [6] [0] [2nd] [2] (L2) [ENTER].

L1	L2	L3
0	.88763	53.258
60	2.5992	155.95
120	3.81	228.6
180	4.5795	274.77
240	4.9672	298.03
300	5.0325	301.95
360	4.835	290.1

L3 =60L2

Return to the home screen. Find the sum of the areas of the rectangles with

TI-82 [2nd] [STAT] (LIST) [▶] (MATH) [5] (SUM) [2nd] [3] (L3) [ENTER].

TI-83 [2nd] [STAT] (LIST) [▶] [▶] (MATH) [5] (SUM) [2nd] [3] (L3) [ENTER].

sum L3
    2573.090369

We estimate, using 12 left rectangles, that 2,573 customers came to the Saturday sale.

Note: The values in Table 6.2 in your text and the final result differ slightly than those in your lists. This is because the unrounded model found with the data was used for computations in the text. If you have the unrounded model available, you should use it instead of the rounded model $c(m)$ that is given above.

6.1.2 APPROXIMATIONS WITH RIGHT RECTANGLES

When using left rectangles to approximate the results of change, the rightmost data point is not the height of a rectangle and is not used in the computation of the left-rectangle area. Similarly, when using right rectangles to approximate the results of change, the *leftmost* data point is not the height of a rectangle and is not used in the computation of the right-rectangle area.

The following data shows the rate of change of the concentration of a drug in the blood stream in terms of the number of days since the drug was administered:

Day	1	5	9	13	17	21	25	29
Concentration ROC (µg/mL/day)	1.5	0.75	0.33	0.20	0.10	¯1.1	¯0.60	¯0.15

First, we fit a piecewise model to the data.

Clear all lists. Enter the days in L1 and the rate of change of concentration in L2.

Draw a scatter plot of the data. It is obvious that a piecewise model should be used with $x = 20$ as the "break" point.

Delete the last three data points from lists L1 and L2 with [DEL]. Enter {21, 25, 29} in L3 and in L4 enter the corresponding outputs: {¯1.1, ¯0.60, ¯0.15}.

Fit an exponential model to the data in L1 and L2 and put that model in Y1.

L2	L3	L4
1.5	21	¯1.1
.75	25	¯.6
.33	29	¯.15
.2	------	
.1		

L4(4)=

TI-82   Fit a linear model to the data in L3 and L4 with the keystrokes $\boxed{\text{STAT}}$ $\boxed{\blacktriangleright}$ (CALC) $\boxed{5}$ (LinReg) $\boxed{\text{2nd}}$ $\boxed{3}$ (L3) $\boxed{,}$ $\boxed{\text{2nd}}$ $\boxed{4}$ (L4) $\boxed{\text{ENTER}}$.

Copy the model to Y2.

```
Y₁◻1.70819479188
27*.844976547832
36^X
Y₂◻.11875X+⁻3.58
54166666667
Y₃=
Y₄=
Y₅=
```

TI-83   Fit a linear model to the data in L3 and L4 and copy it to Y2 with the keystrokes $\boxed{\text{STAT}}$ $\boxed{\blacktriangleright}$ (CALC) $\boxed{4}$ (LinReg) $\boxed{\text{2nd}}$ $\boxed{3}$ (L3) $\boxed{,}$ $\boxed{\text{2nd}}$ $\boxed{4}$ (L4) $\boxed{,}$ $\boxed{\text{VARS}}$ $\boxed{\blacktriangleright}$ (Y-VARS) $\boxed{1}$ (Function) $\boxed{1}$ (Y2) $\boxed{\text{ENTER}}$.

Both (optional)   If you want to graph this model on the original scatter plot, first re-enter the data with all the inputs in list L1, all the outputs in list L2.

Return to the graphing list and turn *off* Y1 and Y2.

Enter (Y1)(X≤20) + (Y2)(X>20) in Y3. (Use the Y-VARS menu to type Y1 and Y2.) Choose Dot mode, draw the scatter plot and Y3 will graph over it.

Y3 = (Y1)(X≤20)+(Y2)(X>20)

Now, we determine the right-rectangle area for $0 \le x \le 20$:

Clear all lists. Because x starts at 0 and $\Delta x = 2$, enter in L1 the values 0, 2, 4, 6, ..., 20 or use seq(X, X, 0, 20, 2) to generate the list. Because we are using right rectangles, *delete* the leftmost input value (0).

Use either the model in Y1 or Y3 to generate the outputs.

L1	L2	L3
2	1.2196	------
4	.8708	
6	.62174	
8	.44391	
10	.31695	
12	.22629	
14	.16157	

L₂=Y₁(L₁)

Each rectangle has width 2. The heights of the rectangles are in list L2. Enter the rectangle areas in list L3 as width*height = 2*L2.

L1	L2	L3
2	1.2196	2.4393
4	.8708	1.7416
6	.62174	1.2435
8	.44391	.88782
10	.31695	.63389
12	.22629	.45259
14	.16157	.32314

L₃=2L₂

Return to the home screen and enter sum L3 with the keystrokes $\boxed{\text{2nd}}$ $\boxed{\text{STAT}}$ (LIST) $\boxed{\blacktriangleright}$ (MATH) $\boxed{5}$ (SUM) $\boxed{\text{2nd}}$ $\boxed{3}$ (L3) $\boxed{\text{ENTER}}$.

We estimate, using 10 right rectangles, that the change in concentration was approximately 8.24 µg/mL.

```
sum L3
        8.234813648
```

To estimate, using right rectangles, the change in drug concentration for $20 \le x \le 29$ days with $\Delta x = 1$, follow the same procedure as above. The values in L1 begin with 21 because we must eliminate the leftmost value (20) when using right rectangles. You should use the absolute value of the model in Y2 (or Y3) to generate L2.

Since the width of each rectangle is 1, L2 contains the areas as well as the heights of the 9 right rectangles.

L1	L2	L3
21	1.0917	------
22	.97292	
23	.85417	
24	.73542	
25	.61667	
26	.49792	
27	.37917	

L₂=abs Y₂(L₁)

sum L2 = 5.55 µg/mL

6.2 Trapezoid and Midpoint Rectangle Approximations

You can compute areas of trapezoids on the home screen of your calculator or use the fact that the trapezoid approximation is the average of the left- and right-rectangle approximations. Areas of midpoint rectangles are found in the same manner as left and right rectangle areas except that the midpoint of the base of each rectangle is in L1 and no data values are deleted. However, such procedures can become tedious when the number, n, of subintervals is large.

6.2.1 SIMPLIFYING AREA APPROXIMATIONS   When you have a model $y = f(x)$ in Y1 in the Y= list, you will find program NUMINTGL very helpful in determining left-rectangle, right-rectangle, midpoint-rectangle, and trapezoidal numerical approximations for accumulated change. Program NUMINTGL is listed in the TI-82/TI-83 Appendix.

WARNING: This program will not work properly if you have any functions in your Y= list that have letters other than X in them. Delete any such functions (like the logistic equation that might be in Y4) before using program NUMINTGL. If you receive an error while running the program, you may have a picture or program stored to a single-letter name. For instance, if program NUMINTGL is trying to store a number in T and you have a program called T, the calculator stops. Delete or rename any programs or pictures that you have called by a single-letter name before continuing.

We illustrate using this program with a model for the Carson River flow rates:

$$f(h) = 18{,}225h^2 - 135{,}334.3h + 2{,}881{,}542.9 \text{ cubic feet per hour}$$

h hours after 11:45 a.m. Wednesday. (The complete model found from the data in your text is used for all the following calculations.)

Have $f(h)$ in Y1. Your function *must* be in Y1 for program NUMINTGL to operate. If you wish to view the approximating rectangles or trapezoids, the program will automatically draw a graph of the function when enter the input interval.	`Y1=18225X^2+ -135` `334.28571428X+28` `81542.8571428` `Y2=` `Y3=` `Y4=` `Y5=` `Y6=`
Start program NUMINTGL by pressing PRGM followed by the number next to the location of the program. Press ENTER. (At this point, if you did not enter the function in Y1, enter 2 to exit the program. Enter Y1 and re-run the program.) If your function is in Y1, enter 1 to continue.	`ENTER F(X) IN Y1` `CONTINUE?` `YES(1) NO(2)`
At the next prompt, press 1 ENTER to draw the approximating figures or press 2 ENTER to obtain only the numerical approximations to the area between $f(h)$ and the horizontal axis between 0 and 20. We choose to see some pictures.	`ENTER F(X) IN Y1` `CONTINUE?` `YES(1) NO(2) 1` `DRAW PICTURES?` `YES(1) NO(2) 1`
At the LEFT ENDPOINT? prompt, type 0 ENTER, and at the RIGHT ENDPOINT? prompt, 20 ENTER to tell the calculator the input interval. You are next shown a menu of choices. Press 1 ENTER to find a left-rectangle approximation.	`ENTER CHOICE:` `LEFT RECT  (1)` `RIGHT RECT (2)` `TRAPEZOIDS (3)` `MIDPT RECT (4) 1`

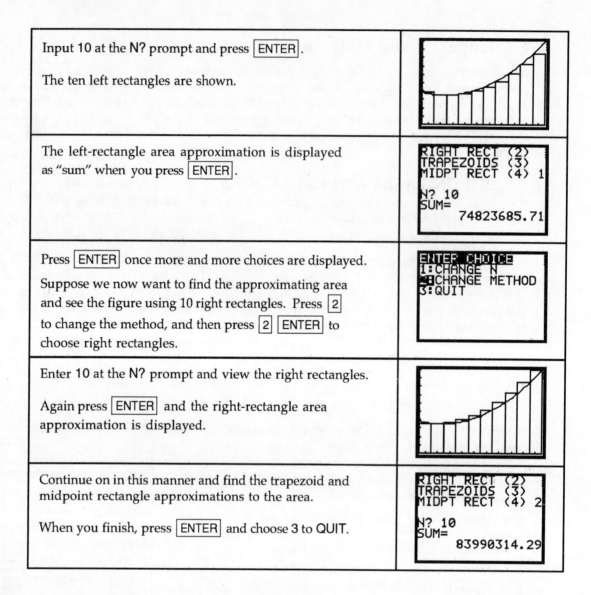

Input 10 at the N? prompt and press ENTER. The ten left rectangles are shown.	
The left-rectangle area approximation is displayed as "sum" when you press ENTER.	RIGHT RECT (2) TRAPEZOIDS (3) MIDPT RECT (4) 1 N? 10 SUM= 74823685.71
Press ENTER once more and more choices are displayed. Suppose we now want to find the approximating area and see the figure using 10 right rectangles. Press 2 to change the method, and then press 2 ENTER to choose right rectangles.	ENTER CHOICE 1:CHANGE N 2:CHANGE METHOD 3:QUIT
Enter 10 at the N? prompt and view the right rectangles. Again press ENTER and the right-rectangle area approximation is displayed.	
Continue on in this manner and find the trapezoid and midpoint rectangle approximations to the area. When you finish, press ENTER and choose 3 to QUIT.	RIGHT RECT (2) TRAPEZOIDS (3) MIDPT RECT (4) 2 N? 10 SUM= 83990314.29

6.3 The Definite Integral as a Limit of Sums

This section introduces you to a very important and useful concept of calculus -- the definite integral. Your calculator can be very helpful as you study definite integrals and how they relate to the accumulation of change.

6.3.1 LIMITS OF SUMS
When you are looking for a trend in midpoint-rectangle approximations to the area between a non-negative function and the horizontal axis between two values of the input variable, program NUMINTGL is extremely useful! However, for this use of the program, it is not advisable to draw pictures when n, the number of subintervals, is large.

To construct a chart of midpoint approximations for the area between $f(x) = \sqrt{1-x^2}$ and the x-axis from $x=0$ and $x=1$, first enter $f(x)$ in Y1. (Don't forget to enclose $1-x^2$ in parentheses.)	Y1=√(1-X²) Y2= Y3= Y4= Y5= Y6= Y7= Y8=

Start program NUMINTGL. Since many subintervals are used with the idea of a limit of sums, do *not* choose to draw the pictures unless you have time to sit and wait for all the rectangles to draw! After entering 0 and 1 as the respective lower and upper limits, choose option 4 for midpoint rectangles.	```ENTER CHOICE:` `LEFT RECT (1)` `RIGHT RECT (2)` `TRAPEZOIDS (3)` `MIDPT RECT (4)` `SIMPSONS (5) 4```
Input some number of subintervals, say N = 4. Record on paper the midpoint area approximation 0.7959823052. (If you want an answer accurate to the thousandths position, record at least 4 decimal places.)	```RIGHT RECT (2)` `TRAPEZOIDS (3)` `MIDPT RECT (4) 4` `N? 4` `SUM=` ` .7959823052```
Press ENTER and choose option 1: CHANGE N. Double the number of subintervals to N = 8. Record the midpoint approximation 0.7891717328 (again, to at least 4 decimal places).	```N? 8` `SUM=` ` .7891717328```
Continue on in this manner, each time choosing option 1: CHANGE N and doubling N until a trend is evident. (Finding a trend means that you can tell what value the approximations are getting closer and closer to, within a specified accuracy, without having to run the program ad infinitum!) Choose QUIT (3) to stop.	```N? 256` `SUM=` ` .7854191806```

- Remember that the trend indicated by the limit of sums can be interpreted as the area of the region between the function and the input axis *only* when that region lies above the input axis. When the region lies below the input axis, the trend is the negative of the area of the region.

6.5 The Fundamental Theorem

Recall that nDeriv is the calculator's numerical derivative and provides, in most cases, a good approximation to the instantaneous rate of change of a function when that rate of change exists. Your calculator can also give you a numerical approximation for a definite integral of a function. This numerical integrator is called fnInt, and the correspondence between the mathematical definite integral notation $\int_{a}^{b} f(x)\, dx$ and the calculator's notation fnInt(f(X), X, a, b) is as shown below:

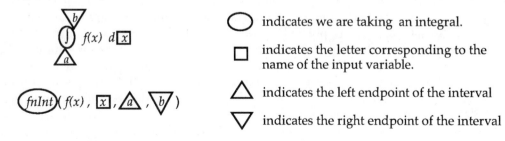

$\bigcirc$ indicates we are taking an integral.

$\square$ indicates the letter corresponding to the name of the input variable.

$\triangle$ indicates the left endpoint of the interval

$\triangledown$ indicates the right endpoint of the interval

You will find that in most cases, your calculator's numerical integrator gives a very close decimal approximation of the exact value of a definite integral. This use is examined in Section 6.6 of this *Guide*. In this section, we illustrate the Fundamental Theorem of Calculus and see how to draw the graph of a general antiderivative of a function.

6.5.1 THE FUNDAMENTAL THEOREM OF CALCULUS   This theorem tells us that the derivative of an antiderivative of a function is the function itself. Let us view this theorem both numerically and graphically.

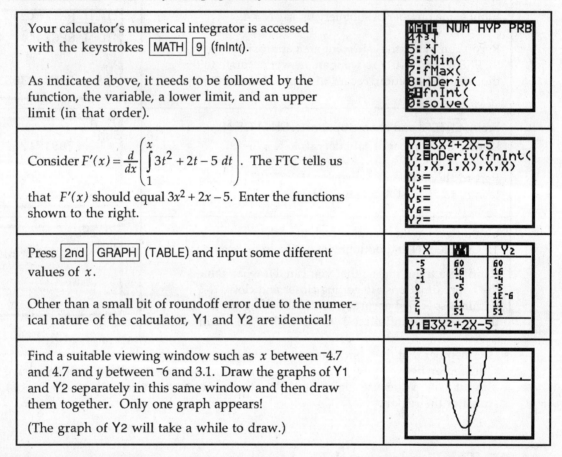

Your calculator's numerical integrator is accessed with the keystrokes [MATH] [9] (fnInt().

As indicated above, it needs to be followed by the function, the variable, a lower limit, and an upper limit (in that order).

Consider $F'(x) = \dfrac{d}{dx}\left(\displaystyle\int_1^x 3t^2 + 2t - 5 \ dt\right)$. The FTC tells us that $F'(x)$ should equal $3x^2 + 2x - 5$. Enter the functions shown to the right.

Press [2nd] [GRAPH] (TABLE) and input some different values of x.

Other than a small bit of roundoff error due to the numerical nature of the calculator, Y1 and Y2 are identical!

Find a suitable viewing window such as x between ‾4.7 and 4.7 and y between ‾6 and 3.1. Draw the graphs of Y1 and Y2 separately in this same window and then draw them together. Only one graph appears!

(The graph of Y2 will take a while to draw.)

Enter several other functions in Y1 and perform the same explorations as above. Confirm your results with derivative and integral formulas. Are you convinced?

6.5.2 DRAWING ANTIDERIVATIVE GRAPHS   All antiderivatives of a specific function differ only by a constant. We explore this idea using the function $f(x) = 3x^2 - 1$ and its antiderivative $F(x) = x^3 - x + C$.

The correct syntax for the calculator's numerical integrator is fnInt(f(X), X, a, b) where $f(x)$ is the function you are integrating, x is the variable of integration, a is the lower limit on the integral, and b is the upper limit on the integral. You do *not* have to use x as the variable unless you are graphing the integral or evaluating it using the calculator's table.

Note that you must supply both a lower limit and an upper limit for fnInt. We can use x for the upper limit, but not both the upper and lower limits. Since we are working with a general antiderivative in this illustration, we do not have the starting point for the accumulation. We therefore just choose some value, say 0, to use as the lower limit to illustrate drawing antiderivative graphs. If you choose a different lower limit, your results will differ from those shown below by a constant.

Enter $f(x)$ in Y1, fnlnt(Y1, X, 0, X) in Y2, and $F(x)$ in Y3, Y4, Y5, Y6, and Y7 (using a different value of C in each location)
 	
You can try different values of C than those shown on the right.
	`Y1日3X²-1`
	`Y2日fnInt(Y1,X,0,`
	`X)`
	`Y3日X^3-X+0`
	`Y4日X^3-X-2`
	`Y5日X^3-X-1`
	`Y6日X^3-X+2`
	`Y7日X^3-X-3`

Find a suitable viewing window and graph all the functions. (Try x between ‾3 and 3 and y between ‾5 and 10.) It seems that the only difference in the graphs (other than the graph of Y1) is that the y-intercept is different. But, isn't C the y-intercept?	
Trace the graphs and then jump between them with 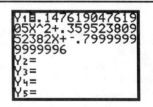. It appears that Y2 and Y3 are the same.	`X=1.2765957  Y=.80386812`

Warning: The methods of Sections 4.4.1 and/or 4.4.2 that we used to check antiderivative formulas are not valid with general antiderivatives. Why? The answer is because in this situation you must choose arbitrarily choose a value for the constant of integration and you must arbitrarily choose a value for the lower limit in order to use the calculator's numerical integrator. However, for most rate-of-change functions where $f(0) = 0$, the calculator's numerical integrator values and your antiderivative formula values should differ by the same constant at every input value where they are defined.

6.6 The Definite Integral

When using the numerical integrator on the home screen, enter fnlnt($f(x)$, x, a, b) for a specific function $f(x)$ with input x and specific values of a and b. (Remember that the input variable does not have to be x on the home screen.) If you prefer, $f(x)$ can be in the Y= list and referred to as Y1 (or whatever location you have it in) when using fnlnt.

6.6.1 EVALUATING A DEFINITE INTEGRAL ON THE HOME SCREEN   We illustrate the use of your calculator's numerical integrator with the model for the rate of change of the average sea level in meters per year during the last 7000 years. A model for these data is $r(t) = 0.14762t^2 + 0.35952t - 0.8$ meters per year t thousand years from the present. (Note that t is negative since we are talking about the past.)

As we have previously mentioned, you should always use the full model (not rounded) for any model you find from data.
 	
Find a quadratic to fit the data shown in the *Changing Sea Levels* example in your text, and enter it in some location of the the Y= list, say Y1.
	`Y1日.147619047619`
	`05X²+.359523809`
	`52382X+‾.7999999`
	`9999996`
	`Y2=`
	`Y3=`
	`Y4=`
	`Y5=`

The model provides a good fit to the data.

Because we are asked for *area* between the input axis and the function, we must find where the function becomes negative.

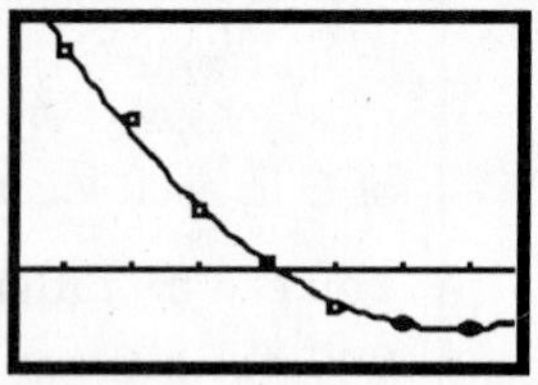

Use the ideas of Section 1.2.3 of this *Guide* to find the *x*-intercept (root or zero) of Y1 between 0 and 7 to be $x \approx -3.845$.

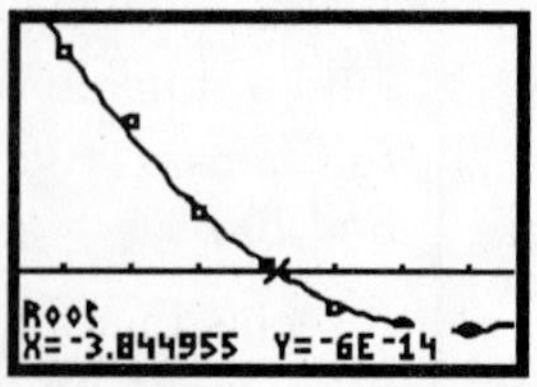

TI-82 Return to the home screen and type the expression shown to the right with the keystrokes [MATH] [9] (fnInt()

[2nd] [VARS] (Y-VARS) [1] (Function) [1] (Y1) [,] [X,T,θ]

[,] [(−)] 7 [,] [(−)] 3.845 [)] .

TI-83 Return to the home screen and type the expression shown to the right with the keystrokes [MATH] [9] (fnInt()

[VARS] [▶] (Y-VARS) [1] (Function) [1] (Y1) [,] [X-T-θ-n]

[,] [(−)] 7 [,] [(−)] 3.845 [)] .

$$\int_{-7}^{-3.845} r(t)\,dt \approx 5.4 \text{ meters}$$

Both The area of the region above the *x*-axis is about 5.4 meters. Now, find the area of the region below the *x*-axis. This area equals the negative of the definite integral of the function over that region.

Evaluate $\int_{-3.845}^{0} r(t)\,dt$ as shown on the right. The area of the region below the *x*-axis is about 2.9 meters.

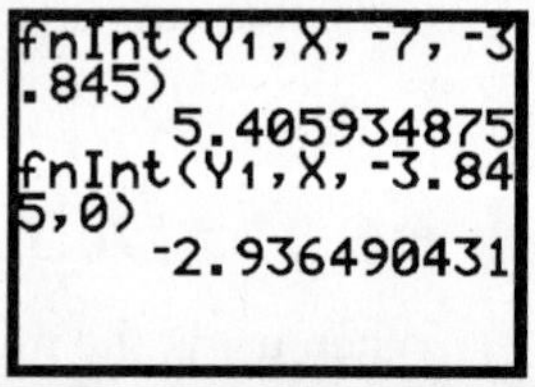

Find $\int_{-7}^{0} r(t)\,dt$.

Note that this value is *not* the sum of the two areas. It is their difference.

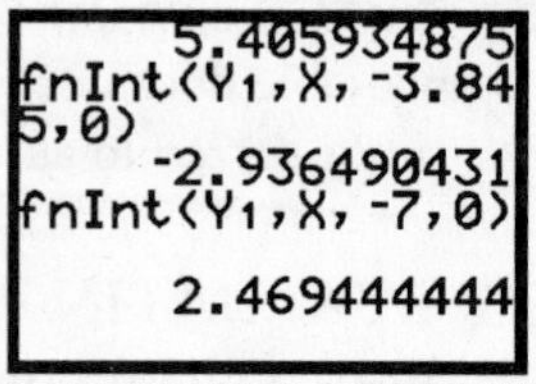

- If you evaluate a definite integral using antiderivative formulas and check your answer with the calculator using fnInt, you may sometimes find a slight difference in the last few decimal places. Remember, the TI-82 and the TI-83 are evaluating the definite integral using an approximation technique.

6.6.2 EVALUATING A DEFINITE INTEGRAL FROM THE GRAPHICS SCREEN

Provided that *a* and *b* are possible *x*-values when you trace the graph, you can find the value of the definite integral $\int_{a}^{b} f(x)\,dx$ from the graphics screen. (For "nice" trace numbers, you can often find the exact *x*-value you need if you graph in the ZDecimal screen or set the viewing window so that Xmax − Xmin equals a multiple of 9.4.)

Turn the STAT PLOTS off, and have the model for the average sea level in Y1 (see Section 6.6.1). Suppose we want to find $\int_{-7}^{0} r(t)\, dt$. Graph Y1.	Y1▪.147619047619 05X^2+.359523809 52382X+-.7999999 9999996 Y2= Y3= Y4= Y5=
Because ⁻7 is not a possible trace value in the window set by the stat plot, press WINDOW and change Xmin to ⁻7 and Xmax to 0. Press 2nd TRACE (CALC) 7 (∫f(x)dx). When the calculator asks for a lower bound, press and hold ◄ until you reach ⁻7.	Lower Limit? X=-7 Y=3.9166667
Press ENTER to set the lower limit and have the calculator ask for the upper limit. Press and hold ► until you reach X = 0.	Upper Limit? X=0 Y=-.8
Press ENTER to calculate the numerical value of the definite integral. Notice that the two regions whose areas were calculated in the first part of Example 1 in the text are shaded.	∫f(x)dx=2.4694444
TI-83 This calculator allows you to input any value of *x* between Xmin and Xmax when finding the value of a definite integral from the graphics screen. You therefore can either follow the above instructions for both calculators or you can do as indicated below. Set the window shown to the right -- the scatter plot window for the average sea level data with Xmax reset to 0.	WINDOW Xmin=-7.6 Xmax=0 Xscl=1 Ymin=-1.816 Ymax=4.616 Yscl=1 Xres=1
Draw the graph of *r(t)* by pressing GRAPH . Press 2nd TRACE (CALC) 7 (∫f(x)dx). When the calculator asks for a lower limit, simply type ⁻7 and then press ENTER .	Y2=.14761904761905X^2+._ Lower Limit? X=-7
When the calculator asks for an upper limit, type 0 and then press ENTER . The value of the integral is displayed and the two regions are shaded. Note that if you had not previously changed Xmax from ⁻0.4 to 0, you would have gotten an ERR: INVALID message at this point because 0 would not have been a value included between Xmin and Xmax.	Y2=.14761904761905X^2+._ Upper Limit? X=0

Chapter 7 Analyzing Accumulated Change: More Applications of Integrals

📖 7.1 Differences of Accumulated Changes

This chapter helps you effectively use your calculator's numerical integrator with various applications pertaining to the accumulation of change. In this first section, we focus on the differences of accumulated changes.

7.1.1 FINDING THE AREA BETWEEN TWO CURVES Finding the area of the region enclosed by two functions uses many of the techniques presented in preceding sections. Suppose we want to find the difference between the accumulated change of $f(x)$ from a to b and the accumulated change of $g(x)$ from a to b where $f(x) = 0.3x^3 - 3.3x^2 + 9.6x + 3.3$ and $g(x) = {}^-0.15x^2 + 2.03x + 3.33$. The input of the leftmost point of intersection of the two curves is a and the input of the rightmost point of intersection is b.

Enter the functions in the Y= list.
 	
First, we must find where the two functions intersect.
	`Y1⊟.3X^3-3.3X²+9`
	`.6X+3.3`
	`Y2⊟-.15X²+2.03X+`
	`3.33`
	`Y3=`
	`Y4=`
	`Y5=`
	`Y6=`

From Figure 7.9 in the text, we see that Xmin = 1, Xmax = 7, Ymin = 0, and Ymax = 12. Press WINDOW , set these values, and then press GRAPH .	
Next, find the two points of intersection. (The x-values of these intersection points will be the limits on the integrals we use to find the area.) Use 2nd TRACE (CALC) 5 (intersect) to locate the intersection point furthest to the left. At the First curve? prompt, press ENTER .	`First curve?` `X=4　　　　Y=8.1`
The cursors jumps to the other curve. Press ENTER . **TI-82** Notice the changing numbers corresponding to the function locations as you move between the curves. **TI-83** Notice the equations corresponding to the functions as you move between the curves. (not shown)	`Second curve?` `X=4　　　　Y=9.05`
Press ◀▓ to move the cursor near the left point of intersection of the two curves. Press ENTER .	`Guess?` `X=3.7446809　Y=8.8283069`

The left point of intersection is displayed. To avoid making a mistake copying the *x*-value and to eliminate as much rounding error as possible, return to the home screen and store this value in *A* with the keystrokes X-T-θ STO▸ ALPHA MATH (A).	Intersection X=3.7150459 Y=8.8013082
Press GRAPH, and repeat the process to find the right intersection point. Do not forget to move the cursor near the intersection point on the right at the Guess? prompt. Store the *x*-value in *B*.	Intersection X=6.7809846 Y=10.198136
The combined area of the three regions is $$\int_1^A (f-g)\,dx + \int_A^B (g-f)\,dx + \int_B^7 (f-g)\,dx.$$ (If you are not sure which curve is on top in each of the various regions, trace the curves.)	fnInt(Y₁-Y₂,X,1,A)+fnInt(Y₂-Y₁,X,A,B)+fnInt(Y₁-Y₂,X,B,7) 17.5132156
Use either the home screen or the graphics screen to find the value of the definite integral $$\int_1^7 (f-g)\,dx = \int_1^7 \big(f(x)-g(x)\big)\,dx.$$ Note that the value of the integral is *not* the same as the area between the two curves.	fnInt(Y₁-Y₂,X,1,7) 2.4

📖 7.3 Streams in Business and Biology

You will find your calculator very helpful when dealing with streams that are accumulated over finite intervals. However, because your calculator's numerical integrator evaluates only definite integrals, you cannot use it to find the value of an improper integral.

7.3.1 FUTURE VALUE OF A DISCRETE INCOME STREAM   We use the sequence command to find the future value of a discrete income stream. The change in the future value at the end of *T* years that occurs because of a deposit of \$*A* at time *t* where interest is earned at an annual rate of 100*r*% compounded *n* times a year is

$$f(t) = A\left(1 + \frac{r}{n}\right)^{n(T-t)} \quad \text{dollars per compounding period}$$

where *t* is the number of years since the first deposit was made. We assume the initial deposit is made at time *t* = 0 and the last deposit is made at time $T - \frac{1}{n}$. The increment for a discrete stream involving *n* compounding periods and deposits of \$*A* at the beginning of each compounding period is $\frac{1}{n}$. Thus, the sequence command for finding the future value of the discrete income stream is

seq(f(X), X, 0, T−1/n, 1/n)

Suppose that you invest \$75 each month in a savings account yielding 6.2% APR compounded monthly. What is the value of your savings in 3 years? To answer this question, note that the change in the future value that occurs due to the deposit at time t is

$$f(t) = 75\left(1 + \frac{0.062}{12}\right)^{12(3-t)} \approx 75\left(1.005166667^{12}\right)^{(3-t)} \approx 75(1.06379)^{(3-t)}$$

Now, find the future value of this stream with $A = 75$, $r = 0.062$, $T = 3$, and $n = 12$.

Enter $f(t)$, using X as the input variable, in location Y1 of the Y= list.
 	
If you want an exact answer, enter the following:
 	
$\quad$ Y1 = 75(1+ 0.062/12)^(12(3–X))
 	
(You must carefully use parentheses with this form of the equation.)
	```Y1■75(1.06379)^(3-X)```
	```Y2=```
	```Y3=```
	```Y4=```
	```Y5=```
	```Y6=```
	```Y7=```

Return to the home screen.
 	
Enter seq(Y1, X, 0, 3 –1/12, 1/12) . Recall that the seq command is accessed with [2nd] [STAT] (LIST) [5] (seq( ).
 	
(Note that since you start counting at 0, the ending value will be one increment less than the number of years the money accumulates in the account.)
	```seq(Y1,X,0,3-1/12,1/12)```
	```(90.28777982 89…```
	
	Press [ENTER].

The sequence is generated. (You can scroll through the list with ■▶.) This list contains the heights of the 36 left rectangles in dollars per month.
 	
Because the input is in years, dividing by 12 and then multiplying by 12 converts the list to the areas of the 36 rectangles in dollars.
 	
The future value we seek is the sum of the heights of the rectangles.
	```seq(Y1,X,0,3-1/12,1/12)```
	```(90.28777982 89…```
	```sum Ans```
	``` 2974.327076```

- The TI-83 can be used in the this manner for sequences with a large number of entries. However, *TI-82 sequences cannot contain more than 99 entries.* In the above example, the list of heights contains 12*3 = 36 entries. However, suppose you needed the future value of this discrete income stream 40 years after the initial deposit is made. The list of heights would contain 12*40 = 480 entries!

   When you try to generate a sequence of more than 99 values with the TI-82, you will get an error message. If this happens, you should break the sequence up into several smaller sequences. For instance, in the example above, you could have entered

   $\quad$ seq(Y1, X, 0, 2 –1/12, 1/12) + seq(Y1, X, 2, 3 –1/12, 1/12)

   to obtain 2043.523226 + 930.8038499 ≈ \$2974.34.

- When using definite integrals to approximate either the future value or the present value of a discrete income stream or to find the future value or the present value of a continuous stream, use the fnInt command to find the area between the appropriate continuous rate of change function and the t-axis from 0 to T.

- The TI-83 calculator has a built-in financial menu accessed with [2nd] [x⁻¹] (FINANCE). The items in this menu can be extremely useful in financial applications. Consult your *Owners' Manual* for more details and instructions for use.

📖 7.4 Integrals in Economics

Consumers' and producers' surplus, being defined by definite integrals, are easy to find using the calculator. You should always draw graphs of the demand and supply functions and think of the surpluses in terms of area to better understand the questions being asked.

7.4.1 CONSUMERS' SURPLUS   Suppose that the demand for mini-vans in the United States can be modeled by $D(p) = 14.12(0.933)^p - 0.25$ million mini-vans when the market price is p thousand dollars per mini-van.

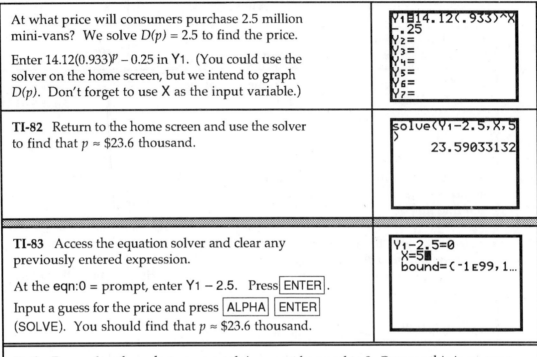

At what price will consumers purchase 2.5 million mini-vans? We solve $D(p) = 2.5$ to find the price. Enter $14.12(0.933)^p - 0.25$ in Y1. (You could use the solver on the home screen, but we intend to graph $D(p)$. Don't forget to use X as the input variable.)	Y1⊟14.12(.933)^X -.25 Y2= Y3= Y4= Y5= Y6= Y7=
TI-82 Return to the home screen and use the solver to find that $p \approx \$23.6$ thousand.	solve(Y1-2.5,X,5) 23.59033132
TI-83 Access the equation solver and clear any previously entered expression. At the **eqn:0 =** prompt, enter Y1 − 2.5. Press ENTER . Input a guess for the price and press ALPHA ENTER (SOLVE). You should find that $p \approx \$23.6$ thousand.	Y1-2.5=0 X=5■ bound=(-1E99,1...

Both   Remember that what you are solving must be equal to 0. Because this is an exponential equation, it can have no more than one solution. Therefore, your guess can be any reasonable value. Refer to Section 1.2.2 of this *Guide* for instructions on using the solver.)

Now, let us find if the model indicates a possible price above which consumers will purchase no mini-vans. If so, we will find that price.

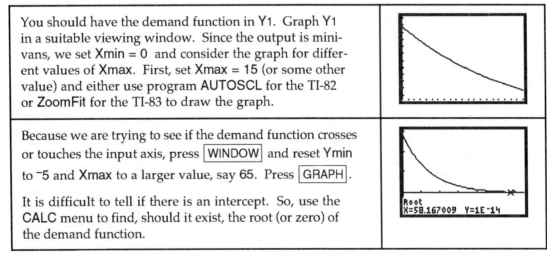

You should have the demand function in Y1. Graph Y1 in a suitable viewing window. Since the output is mini-vans, we set Xmin = 0 and consider the graph for different values of Xmax. First, set Xmax = 15 (or some other value) and either use program AUTOSCL for the TI-82 or ZoomFit for the TI-83 to draw the graph.	
Because we are trying to see if the demand function crosses or touches the input axis, press WINDOW and reset Ymin to ⁻5 and Xmax to a larger value, say 65. Press GRAPH . It is difficult to tell if there is an intercept. So, use the CALC menu to find, should it exist, the root (or zero) of the demand function.	Root X=58.167009 Y=1E⁻14

Instead of drawing the graph, you could have used the solver to find the location of the x-intercept. Notice that here we are looking for where Y1 = 0. (Use the equation solver on the TI-83.) Consumers will not pay more than about $58,200 per mini-van.	```solve(Y1,X,5)``` ``` 58.16700863```
What is the consumers' surplus when 2.5 million mini-vans are purchased? We earlier found that the market price for this quantity is 23.59033 thousand dollars. The consumers' surplus is *approximately* the area of the region shown to the right (the trace value on the TI-82 is not set exactly to the market price).	```∫f(x)dx=27.59539```

Find the consumers' surplus, which is the area of the shaded region, by finding the value of
 	
$$\int_{M}^{P} D(p)\, dp = \int_{23.59033132}^{58.16700863} \left(14.12(0.933)^p\right) dp.$$
 	
(Remember that you can avoid retyping the long decimal numbers by storing them to various memory locations as you find them.)
	```M```
	``` 23.59033132```
	```P```
	``` 58.16700863```
	```fnInt(Y1,X,M,P)```
	``` 27.40481673```

- Carefully watch the units involved in your computations. Refer to the statement of the problem and note that the height is measured in thousand dollars per mini-van and the width in million mini-vans. Thus, the area should be written in units that make sense in the context of this problem:

$$\text{height} * \text{width} \approx 27.4 \left(\frac{\text{thousand dollars}}{\text{minivan}}\right)(\text{million minivans}) = \$\,27.4 \text{ thousand million}$$

$$= \$27.4\,(1{,}000)(1{,}000{,}000) = \$27.4(10^9) = \$27.4 \text{ billion dollars}$$

7.4.2 PRODUCERS' SURPLUS   When dealing with supply functions, use definite integrals in a manner similar to that for consumers' quantities to find producers' revenue, producers' surplus, and so forth. To illustrate, suppose the demand for mini-vans in the United States can be modeled by $D(p) = 14.12(0.933)^p - 0.25$ million mini-vans and the supply curve is $S(p) = 0$ million mini-vans when $p < 15$ and $S(p) = 0.25p - 3.75$ million mini-vans when $p \geq 15$ where the market price is p thousand dollars per mini-van.

Enter the demand curve in Y1 and the supply curve in Y2. (Remember to use X as the input variable.)
 	
Draw a graph of the demand and supply functions in an appropriate window, say $0 \leq x \leq 65$ and $-5 \leq y \leq 16$.
	```Y1■14.12(.933)^X```
	```-.25```
	```Y2■0(X<15)+(.25X```
	```-3.75)(X≥15)```
	```Y3=```
	```Y4=```
	```Y5=```
	```Y6=```

Market equilibrium occurs when $D(p) = S(p)$. Use the methods of Section 7.1.1 of this *Guide* to find the intersection of the two curves. Store the x-value of the intersection as B.	```Intersection``` ```X=24.399631 Y=2.3499077```

Producers' surplus is found by evaluating $\int_{15}^{B} S(p)\, dp$

```
X→B
         24.39963071
fnInt(Y₂,X,15,B)
         11.04413219
```

7.5 Average Value and Average Rates of Change

You need to carefully read any question involving average value in order to determine which quantity is involved. Considering the units of measure in the situation can be of tremendous help when trying to determine which function to integrate when finding average value.

7.5.1 AVERAGE VALUE OF A FUNCTION

Suppose that the hourly temperatures shown below were recorded from 7 a.m. to 7 p.m. one day in September.

Time	7am	8	9	10	11	noon	1pm	2	3	4	5	6	7
Temp. (°F)	49	54	58	66	72	76	79	80	80	78	74	69	62

Enter the input data in L1 as the number of hours after midnight: 7 am is 7 and 1 pm is 13, etc.

First, we fit a cubic model $t(h)$ to the data where h is the number of hours after midnight. The output of the model is measured in degrees Fahrenheit.

Graph the model on the scatter plot of the data and see that it provides a good fit.

```
CubicReg
y=ax³+bx²+cx+d
a=-.0352564103
b=.7181568432
c=1.584332334
d=13.68931069
```

Next, you are asked to calculate the average temperature. Because temperature is measured in this example in degrees Fahrenheit, the units on your result should be °F. When evaluating integrals, it helps to think of the units of the integration result as (height)(width) where the height units are the output units and the width units are the input units of the function that you are integrating. That is,

$$\int_{9 \text{ hours}}^{18 \text{ hours}} t(h) \text{ degrees } dh \quad \text{has units of (degrees)(hours).}$$

When we find the average value, we divide the integral by (upper limit − lower limit). So,

$$\text{average value} = \frac{\displaystyle\int_{9 \text{ hours}}^{18 \text{ hours}} t(h) \text{ degrees } dh}{18 \text{ hours} - 9 \text{ hours}} = \frac{(T(18) - T(9)) \text{ degrees} \cdot \text{hours}}{9 \text{ hours}}$$

where $T(h)$ is an antiderivative of $t(h)$. Because the "hours" cancel, the result is in degrees as is desired. Remember, when finding average value, *the units of the average value are always the same as the output units of the quantity you are integrating.*

Find the average value of the temperature between 9 a.m. and 6 p.m. to be approximately 74.4 °F.

```
fnInt(Y₁,X,9,18)
        669.8450612
Ans/(18-9)
        74.42722902
```

The third part of this example asks you to find the average rate of change of temperature from 9 a.m. to 6 p.m. Again, let the units be your guide. Because temperature is measured in °F and input is measured in hours, the average rate of change is measured in output units per input units = °F per hour. Thus, we find the average rate of change to be

$$\text{average rate of change of temperature} = \frac{(t(18) - t(9))\ ^\circ F}{(18 - 9)\ \text{hours}} = 0.98\ ^\circ F \text{ per hour}$$

7.5.2 GEOMETRIC INTERPRETATION OF AVERAGE VALUE

What does the average value of a function mean in terms of the graph of the function? Consider the model we found above for the temperature one day in September between 7 a.m. and 7 p.m.:

$$t(h) = -0.03526h^3 + 0.71816h^2 + 1.584h + 13.689 \text{ degrees } t \text{ hours after 7 a.m.}$$

You should already have the unrounded model for *t(h)* in Y1. If not enter it now. Turn off the scatter plots and reset the viewing window to that given on the right. Notice the input is between 9 a.m. and 6 p.m.	WINDOW FORMAT Xmin=9 Xmax=18 Xscl=25 Ymin=45.9 Ymax=83.1 Yscl=100
Enter the average value in location Y2 of the Y= list.	Y1=-.03525641025 642X^3+.71815684 315724X^2+1.5843 323343274X+13.68 9310689335 Y2=74.42722902 Y3= Y4=
Press GRAPH. Notice that the area of the rectangle whose height is the average temperature is (74.42722902)(18 − 9) ≈ 669.8.	

The area of the rectangle equals the area of the region between the temperature function *t(h)* and the *t*-axis between 9 a.m. and 6 p.m.
 	
(In this application, the area does not have a meaningful interpretation because its units are (degrees)(hours).)
	74.42722902*9
	   669.8450612
	fnInt(Y1,X,9,18)
	   669.8450612

📖 7.6 Probability Distributions and Density Functions

Most of the applications of probability distributions and density functions use technology techniques that have already been discussed. Probabilities are areas whose values can be found by integrating the appropriate density function. A cumulative density function is an accumulation function of a probability density function.

You will find your calculator's numerical integrator especially useful when finding means and standard deviations for some probability distributions because those integrals often contain expressions for which we have not developed an algebraic technique for finding an antiderivative.

The TI-83 calculator contains many built-in statistical menus that deal with probability distributions and density functions. These are especially useful when you take a statistics course, but some can be used with Section 7.6 in this course. Consult your TI-83 *Owners' Manual* for more details and instructions for use.

7.6.1 NORMAL PROBABILITIES   The normal density function is the most well known and widely used probability distribution. If you are told that a random variable x has a normal distribution $N(x)$ with mean μ and standard deviation σ, the probability that x is between a and b is

$$\int_a^b N(x)\, dx = \int_a^b \frac{1}{\sigma\sqrt{2\pi}} e^{\frac{-(x-\mu)^2}{2\sigma^2}}\, dx\,.$$

Suppose a light-bulb manufacturer advertises that the average life of the company's soft-light bulb is 900 hours with a standard deviation of 100 hours. Suppose also that we know the distribution of the life of these bulbs, with the life span measured in hundreds of hours, is modeled by a normal density function. Find the probability that any one of these light bulbs last between 900 and 1000 hours.

We are told that $\mu = 9$ and $\sigma = 1$. (Note that x is measured in hundreds of hours, so we must convert to these units.) Enter $N(x)$ in Y1 using $\mu = 9$ and $\sigma = 1$. (Carefully watch your use of parentheses.)	```Y1B(1/J(2π))e^-((X-9)²/2) Y2= Y3= Y4= Y5= Y6= Y7=```
Find $p(9 < x < 10)$. (Remember that any probability must be between 0 and 1.)	```fnInt(Y₁,X,9,10) .3413447461```
To draw a graph of the normal density function, note that nearly all of the area between the function and the horizontal axis lies within 3 standard deviations of the mean. Set Xmin = 9 −3 = 6 and Xmax = 9 + 3 = 12. Use **AUTOSCL** on the TI-82 or **ZoomFit** on the TI-83 to draw the graph.	

We could have graphically found $p(9 < x < 10)$ using the **CALC** menu. On the TI-82, you need the horizontal view is chosen so that 9 and 10 are possible trace values. Remember that the width of the screen must be a multiple of 9.4 to trace with "nice" numbers. Set Xmin = 9 − 9.4/2 = 4.3 and Xmax = 9 + 9.4/2 = 13.7. Since you can type the upper and lower bounds on the TI-83, you do not need to change the window.

Press 2nd TRACE (CALC) 7 (∫f(x)dx), choose **9** as the **Lower Limit**, choose **10** as the **Upper Limit**, and see the shaded region whose area is $p(9 < x < 10)$.	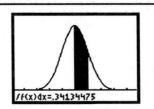

Chapter 8 Repetitive Change: Cycles and Trigonometry

📖 8.1 Functions on Angles: Sines and Cosines

Before you begin this chapter, go back to the first page of this *Guide* and check the basic setup, the statistical setup, and the window setup. If these are not set as specified in Figures 1, 2, and 3, you will have trouble using your calculator in this chapter. Pay careful attention to the third line in the MODE screen in the basic setup. The Radian/Degree mode setting affects the calculator's interpretation of the ANGLE menu choices. Your calculator's MODE menu should always be set to Radian unless otherwise specified.

8.1.1 CONVERTING ANGLES FROM RADIANS TO DEGREES

Even though the process of converting angles from one measure to another is a simple arithmetic task, we cover it here to show you more of your calculator's functionality. Suppose you want to convert $7\pi/6$ to degree measure. (Remember that when no units are specified, the angle is assumed to be expressed in radian measure.)

As shown in the text example, convert $7\pi/6$ radians to degrees with multiplication by the proper factor. (Let the units be your guide and carefully use parentheses.) $$\left(\frac{7\pi}{6}\text{ radians}\right)\left(\frac{180^{\circ}}{\pi\text{ radians}}\right)=210^{\circ}$$	`(7π/6)(180/π)` `                210`

If you want to use your calculator for this conversion *to* degrees, first set the calculator to degree mode. Press MODE, choose Degree in the third line and press ENTER. Return to the home screen.	`Normal Sci Eng`
	`Float 0123456789`
	`Radian Degree`
	`Func Par Pol Seq`
	`Connected Dot`
	`Sequential Simul`
	`Fullscreen Split`

Tell the calculator the angle and that it is measured in radians with (7 2nd ^ (π) ÷ 6) 2nd MATRX (ANGLE) 3 (ʳ). Press ENTER to convert the angle to degrees.	`(7π/6)ʳ` `                210`

While it is not necessary in this example, if you have an angle that you want expressed in degrees, minutes, and seconds, first convert the angle to degrees and then press 2nd (–) (ANS) 2nd MATRX (ANGLE) 4 (▶DMS).
 	
(Don't forget to set your calculator mode back to radian measure when you finish the conversions!)
	`(7π/6)ʳ`
	`                210`
	`Ans▶DMS`
	`          210°0'0"`

8.1.2 CONVERTING ANGLES FROM DEGREES TO RADIANS

Converting angles from degree measure to radian measure is similar to process described above. However, since you wish to convert to radian measure, the mode setting should be Radian. Suppose that you need to convert 135° to radian measure.

As shown in the text example, do this numerically by multiplication of the proper factor. (Let the units be your guide and carefully use parentheses.) $$\left(135^{o}\right)\left(\frac{\pi \text{ radians}}{180^{o}}\right) = \frac{3\pi}{4} \text{ radians} = \frac{3\pi}{4}$$	```135(π/180)⎕ 2.35619449⎕3π/4 2.35619449```
Because you want to convert *to* radians, press MODE, choose **Radian** in the third line and press ENTER. Enter the degree measure of the angle and tell the calculator the angle is in degrees with 2nd MATRX (ANGLE) 1 (°). Press ENTER to convert to radians.	```135° 2.35619449```

8.1.3 EVALUATING TRIGONOMETRIC FUNCTIONS

Just as it was important to have the correct mode set when changing from degrees to radians or vice versa, it is essential that you have the correct mode set when evaluating trigonometric function values. Unless you see the degree symbol, the mode setting should be **Radian**.

Find $\sin \frac{9\pi}{8}$ and $\cos \frac{9\pi}{8}$. Because these angles are in radians, be certain that **Radian** is chosen in the third line of the **Mode** screen. The SIN and COS keys are above the , and (keys on your calculator's keyboard.	```sin (9π/8) -.3826834324cos (9π/8) -.9238795325```
It is also essential that you use parentheses -- the calculator's order of operations are such that the sine function takes precedence over division. Without telling your calculator to first divide, it evaluates $\frac{\sin (9\pi)}{8}$ and $\frac{\cos (9\pi)}{8}$.	``` -.3826834324cos (9π/8) -.9238795325sin 9π/8 2.5ᴇ-14cos 9π/8 -.125```

- When you press SIN and/or COS on the TI-83, the left parentheses automatically appears and cannot be deleted. The right parentheses is not necessary, but if you want to use correct mathematical notation, it should be there.

The values of the trigonometric functions are the same regardless of whether the angles are expressed in radian measure or degree measure. To verify this, first convert the angle in radians to degree measure, and then re-evaluate the sine and cosine of the angle.

Change the mode setting to **Degree** and then convert $\frac{9\pi}{8}$ to degrees. (When you finish, don't forget to change the mode setting back to **Radian**!)	```(9π/8)ʳ 202.5sin 202.5 -.3826834324cos 202.5 -.9238795325```

📖 8.2 Cyclic Functions as Models

We now introduce another model -- the sine model. As you might expect, this model should be fit to data that repeatedly varies between alternate extremes. The form of the model is $f(x) = a \sin(bx + c) + d$ where a is the amplitude, $2\pi / |b|$ is the period, $^-c/b$ is the horizontal shift, and d is the vertical shift.

8.2.1 FITTING A SINE MODEL TO DATA

Before fitting any model to data, remember that you should construct a scatter plot of the data and observe what pattern the data appears to follow. We illustrate finding a sine model for cyclic data with the hours of daylight on the Arctic Circle as a function of the day of the year on which the hours of daylight are measured. (January 1 is day 1.)

Day of the year	$^-10$	81.5	173	264	355	446.5	538	629	720	811.5
Hours of daylight	0	12	24	12	0	12	24	12	0	12

Enter these data with the day of the year in L1 and the hours of daylight in L2.

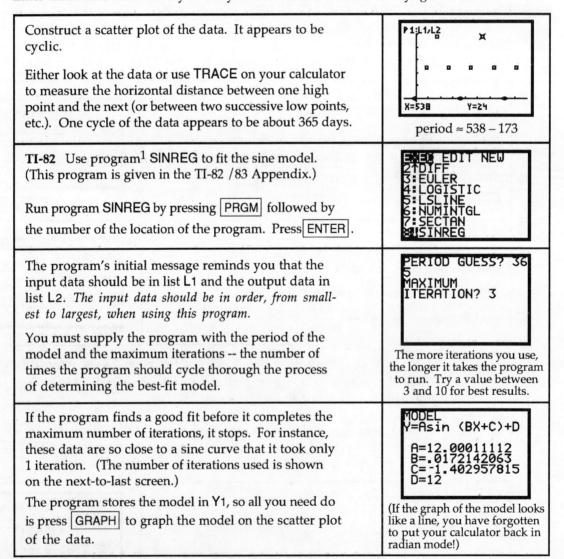

Construct a scatter plot of the data. It appears to be cyclic. Either look at the data or use **TRACE** on your calculator to measure the horizontal distance between one high point and the next (or between two successive low points, etc.). One cycle of the data appears to be about 365 days.	P1:L1,L2 X=538 Y=24 period ≈ 538 − 173
TI-82 Use program[1] SINREG to fit the sine model. (This program is given in the TI-82 /83 Appendix.) Run program SINREG by pressing PRGM followed by the number of the location of the program. Press ENTER.	EXEC EDIT NEW 2↑DIFF 3:EULER 4:LOGISTIC 5:LSLINE 6:NUMINTGL 7:SECTAN 8↓SINREG
The program's initial message reminds you that the input data should be in list L1 and the output data in list L2. *The input data should be in order, from smallest to largest, when using this program.* You must supply the program with the period of the model and the maximum iterations -- the number of times the program should cycle thorough the process of determining the best-fit model.	PERIOD GUESS? 365 MAXIMUM ITERATION? 3 The more iterations you use, the longer it takes the program to run. Try a value between 3 and 10 for best results.
If the program finds a good fit before it completes the maximum number of iterations, it stops. For instance, these data are so close to a sine curve that it took only 1 iteration. (The number of iterations used is shown on the next-to-last screen.) The program stores the model in Y1, so all you need do is press GRAPH to graph the model on the scatter plot of the data.	MODEL Y=Asin (BX+C)+D A=12.00011112 B=.0172142063 C=-1.402957815 D=12 (If the graph of the model looks like a line, you have forgotten to put your calculator back in radian mode!)

[1]This program is based on a program that was modified by John Kenelly from material by Charles Scarborough of Texas Instruments. The authors sincerely thank Robert Simms for this help with this version of the program.

TI-83 The sine model on the TI-83 is in the list with the other models. Press [STAT] [▶] (CALC) [▲] [C] (SinReg). Enter the list containing the input data, the list containing the output data, and the location of the Y= list that you wish the model to appear. Press [ENTER].	`SinReg L₁,L₂,Y₁`

The sine model of best fit appears on the screen and is pasted into the Y1 location of the Y= list.
 	
(If you wish, you can tell the calculator how many times to go through the routine that fits the model. This number of iterations is 3 if not specified. The number should be typed before L1 when initially finding the model.)
	`SinReg`
	`y=a*sin(bx+c)+d`
	`a=12.00011112`
	`b=.0172142063`
	`c=-1.402957815`
	`d=12`

Graph the model on the scatter plot of the data. (If the graph of the model looks like a line, you have forgotten to put your calculator back in radian mode!)	
Even though it did not occur in this example, you may get a **SINGULAR MATRIX** error when trying to fit a sine model to certain data. If so, try specifying an estimate for the period of the model. Recall that our estimate of the period is 365 days.	`SinReg L₁,L₂,365` `,Y₁`

Notice how much faster the TI-83 finds the model!
 	
(If you do not think the original model the calculator finds fits the data very well, try specifying a period and see if a better-fitting model results. It didn't here, but it might with a different set of data.)
	`SinReg`
	`y=a*sin(bx+c)+d`
	`a=12.00011112`
	`b=.0172142063`
	`c=-1.402957815`
	`d=12`

8.3 Rates of Change and Derivatives

All the previous techniques given for other models also hold for the sine model. You can find intersections, maxima, minima, inflection points, derivatives, integrals, and so forth.

8.3.1 DERIVATIVES OF SINE AND COSINE MODELS   Evaluate nDeriv at a particular input to find the value of the derivative of the sine model at that input. Suppose the calls for service made to a county sheriff's department in a certain rural/suburban county can be modeled as $c(h) = 2.8 \sin(0.262h + 2.5) + 5.38$ calls during the hth hour after midnight.

Enter the model in some location of the Y= list, say Y1.
 	
In another location, say Y2, enter the calculator's numerical derivative.
	`Y₁=2.8sin (.262X`
	`+2.5)+5.38`
	`Y₂=nDeriv(Y₁,X,X`
	`)`
	`Y₃=`
	`Y₄=`
	`Y₅=`
	`Y₆=`

TI-83 Because the TI-83 automatically inserts a left parentheses when you press $\boxed{\text{SIN}}$ and $\boxed{\text{COS}}$, be very careful that you do not type an extra parentheses when you type a model in the graphing list. If you do, you will change the order of operations and not have the correct model. The model in Y3 is <u>incorrect</u>.

```
Plot1 Plot2 Plot3
\Y1⊟2.8sin(.262X
+2.5)+5.38
\Y2=nDeriv(Y1,X,
X)
\Y3⊟2.8sin((.262
X+2.5)+5.38
\Y4=
```

When these two models that only differ by the one left parentheses are graphed, you can see that they are entirely different functions! The correct model is the darker of the two.

This warning also applies for the cosine function.

Both Even though this calculator will not give you an equation for the derivative, you can use the methods illustrated in Sections 4.4.1 and/or 4.4.2 of this *Guide* to check the derivative you obtain using derivative rules.

Enter your derivative formula in Y3, turn off Y1, and use the table to compare several values of the calculator's derivative (Y2) and your derivative (Y3).

X	Y2	Y3
0	-.5877	-.5877
2	-.7285	-.7285
4	-.6738	-.6738
8	-.0852	-.0852
15	.72571	.72571
23	-.4567	-.4567

X=

Y3 =
0.7336 cos(0.262X + 2.5)

How quickly is the number of calls received each hour changing at noon? at midnight?

To answer these questions, simply evaluate Y2 (or your derivative in Y3) at 12 for noon and 0 (or 24) for midnight.

(You weren't told if "midnight" refers to the initial time or 24 hours after that initial time.)

```
Y2(12)
          .588774165
Y2(0)
         -.5877189497
Y2(24)
         -.5898259682
```

📖 8.5 Accumulation in Cycles

As with the other models we have studied, applications of accumulated change with the sine and cosine models involve the calculator's numerical integrator fnInt.

8.5.1 INTEGRALS OF SINE AND COSINE MODELS   Suppose that the rate of change of the temperature in Philadelphia on August 27, 1993 can be modeled as

$$t(h) = 2.733 \cos(0.285h - 2.93) \text{ °F per hour}$$

h hours after midnight. Find the accumulated change in the temperature between 9 a.m. and 3 p.m. on August 27, 1993.

You could enter the model in the Y= list. However, since we are not asked to draw a graph, we choose to find the result using the calculator's numerical integrator and the home screen.

The temperature increased by approximately 13 °F.

```
fnInt(2.733cos (
.285H-2.93),H,9,
15)
         12.76901162
```

Chapter 9   Ingredients of Multivariable Change: Models, Graphs, Rates

📖 9.1 Cross-Sectional Models and Multivariable Functions

For a multivariable function with two input variables described by data given in a table with the values of the *first* input variable listed *horizontally* across the top of the table and the values of the *second* input variable listed *vertically* down the left side of the table, obtain cross-sectional models by entering the appropriate data in lists L1 and L2 and then proceeding as indicated in Chapter 1 of this *Guide*.

9.1.1 FINDING CROSS-SECTIONAL MODELS (holding the first input variable constant)
Using the elevation data in Table 9.1 of the text, find the cross-sectional model $E(0.8, n)$ as described below. Remember that "rows" go from left to right horizontally and "columns" go from top to bottom vertically.

When considering $E(0.8, n)$, e is constant at 0.8 and n varies. So, enter the values for n appearing on the left side of the table (vertically) in L1 and the elevations E obtained from the table in the $e = 0.8$ column in L2.	L1 L2 L3 1.5 797.6 1.4 798.1 1.3 798.5 1.2 798.9 1.1 799.2 1 799.5 .9 799.7 L2(1)=797.6
Clear previously-entered functions from the Y= list, and turn Plot1 on. Draw a scatter plot of the $E(0.8, n)$ data with ZOOM 9 (ZoomStat).	*(scatter plot)*
The data appear to be quadratic. Fit a quadratic model and copy the model to the Y= list. (Refer to Section 2.3.2 of this *Guide*.)	Y1=-2.4999999999 999X^2+2.4970588 235293X+799.4897 0588235▮ Y2= Y3= Y4= Y5=
Overdraw the graph of the model on the scatter plot with GRAPH. Because we will find several different cross-sectional models, calling different variables by the same names x and y would be very confusing. It is very important that you call the variables by the names they have been assigned in the application. (See below.)	*(graph with curve)* $E(0.8, n) =$ $-2.5n^2 + 2.497n + 799.490$

- Remember that when finding or graphing a model, your calculator always calls the input variable X and the output variable Y. When working with multivariable functions, you must translate the calculator's model Y1 ≈ $-2.5X^2 + 2.497X + 799.490$ to the symbols used in this problem to obtain $E(0.8, n) = -2.5n^2 + 2.497n + 799.490$.

9.1.2 FINDING CROSS-SECTIONAL MODELS (holding the second input variable constant)

Using the elevation data in Table 9.1 of the text, find the cross-sectional model $E(e, 0.6)$ as discussed below.

When considering $E(e, 0.6)$, n is constant at 0.6 and e varies. Thus, enter the values for e appearing across the top of the table (horizontally) in L1 and the elevations E obtained from the table in the $n = 0.6$ row in L2.	L1 L2 L3 0 802.8 .1 801.6 .2 800.8 .3 800.3 .4 800 .5 799.9 .6 799.9 L2(1)=802.8
You can either turn Y1 "off" by placing the blinking cursor over the equals sign and pressing ENTER or you can delete it. Next, draw a scatter plot of the $E(e, 0.6)$ data.	
The data appears to be cubic. Fit a cubic model and copy it to the Y2 location of the Y= list. Be certain that you translate the calculator's model $Y2 \approx {}^{-}10.124X^3 + 21.347X^2 - 13.972X + 802.809$ to the symbols used in the example. (See the note at the end of Section 9.1.1.)	999X^2+2.4970588 235293X+799.4897 0588235 Y2■-10.124103622 553X^3+21.347160 321612X^2+-13.97 2329485096X+802. 80939112487■
Write this cross-sectional model as $E(e, 0.6) = {}^{-}10.124e^3 + 21.347e^2 - 13.972e + 802.809$ Overdraw the graph of the model on the scatter plot with GRAPH.	

9.1.3 EVALUATING OUTPUTS OF MULTIVARIABLE FUNCTIONS

Outputs of multi-variable functions are found by evaluating the function at the given values of the input variables. One way to find multivariable function outputs is to evaluate them on the home screen. For instance, the compound interest formula tells us that

$$A(P, t) = P\left(1 + \frac{0.06}{4}\right)^{4t} \text{ dollars}$$

is the amount to which P dollars will grow over t years in an account paying 6% interest compounded quarterly. Let's review how to find $A(5300, 10)$.

Type, on the home screen, the formula for $A(P, t)$, substituting in $P = 5300$ and $t = 10$. Press ENTER. Again, be warned that you must carefully use the correct placement of parentheses. Check your result with the values in Table 9.4 in the text to see if \$9614.30 is a reasonable amount.	5300(1+.06/4)^(4 *10) 9614.297566

Even though it is not necessary in this example, you may encounter activities in this section in which you need to evaluate a multivariable function at several different inputs. Instead of individually typing each one on the home screen, there are easier methods. You will also use the techniques shown below in later sections of this chapter.

When evaluating a multivariable function at several different input values, you may find it more convenient to enter the multivariable function in the graphing list. It is very important to note at this point that while we have previously used X as the input variable when entering functions in the Y= list, we do *not* follow this rule when *evaluating* functions with more than one input variable.

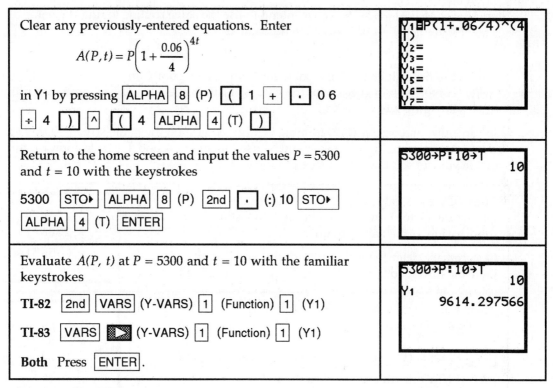

Clear any previously-entered equations. Enter $$A(P,t) = P\left(1 + \frac{0.06}{4}\right)^{4t}$$ in Y1 by pressing $\boxed{\text{ALPHA}}$ $\boxed{8}$ (P) $\boxed{(}$ 1 $\boxed{+}$ $\boxed{.}$ 0 6 $\boxed{\div}$ 4 $\boxed{)}$ $\boxed{\wedge}$ $\boxed{(}$ 4 $\boxed{\text{ALPHA}}$ $\boxed{4}$ (T) $\boxed{)}$	`Y1=P(1+.06/4)^(4` `T)` `Y2=` `Y3=` `Y4=` `Y5=` `Y6=` `Y7=`
Return to the home screen and input the values $P = 5300$ and $t = 10$ with the keystrokes 5300 $\boxed{\text{STO▸}}$ $\boxed{\text{ALPHA}}$ $\boxed{8}$ (P) $\boxed{\text{2nd}}$ $\boxed{.}$ (:) 10 $\boxed{\text{STO▸}}$ $\boxed{\text{ALPHA}}$ $\boxed{4}$ (T) $\boxed{\text{ENTER}}$	`5300→P:10→T` `           10`
Evaluate $A(P, t)$ at $P = 5300$ and $t = 10$ with the familiar keystrokes **TI-82** $\boxed{\text{2nd}}$ $\boxed{\text{VARS}}$ (Y-VARS) $\boxed{1}$ (Function) $\boxed{1}$ (Y1) **TI-83** $\boxed{\text{VARS}}$ $\boxed{▶}$ (Y-VARS) $\boxed{1}$ (Function) $\boxed{1}$ (Y1) **Both** Press $\boxed{\text{ENTER}}$.	`5300→P:10→T` `           10` `Y1` `   9614.297566`

To find another output for this function, say $A(8700, 8.25)$, repeat the above procedure:

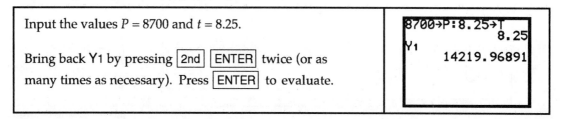

Input the values $P = 8700$ and $t = 8.25$. Bring back Y1 by pressing $\boxed{\text{2nd}}$ $\boxed{\text{ENTER}}$ twice (or as many times as necessary). Press $\boxed{\text{ENTER}}$ to evaluate.	`8700→P:8.25→T` `         8.25` `Y1` `   14219.96891`

Warning: Recall that the your calculator recognizes only X as an input variable in the Y= list when you *graph* and when you use the *table*. Thus, you should not attempt to graph the current Y1 = $A(P, t)$ or use the table because the calculator considers this Y1 a constant. (Check it out and see that the graph is a straight line at about 14220 and that all values in the table are approximately 14220.)

9.2 Contour Graphs

Because any program that we might use to graph[1] a three-dimensional function would be very involved to use and take a long time to execute, we do not attempt to do such. Instead, we discover graphical information about three-dimensional graphs using their associated contour curves.

9.2.1 SKETCHING CONTOUR CURVES

When you are given a multivariable function model, you can draw contour graphs for functions with two input variables using the three-step process described below. We illustrate by drawing the $520 constant-contour curve with the function for the monthly payment on a loan of A thousand dollars for a period of t years:

$$M(t, A) = \frac{5.833333A}{1 - 0.932583^t} \text{ dollars}$$

Step 1: Set $M(t, A) = 520$. Since we intend to use the solve routine to find the value of A at various values of t, write the function as

$$M(t, A) - 520 = 0.$$

Step 2: Choose values for t and solve for A to obtain points on the $520 constant-contour curve. Obtain guesses for the values of t from the table of loan amounts in the table in Figure 9.14 of the text.

You can either enter the $M(t, A)$ formula in Y1 or enter it on the home screen. If you enter the formula in the graphing list, be certain to use the letters A and T and to indicate for which variable you are solving when using the solve routine.

TI-82 Choose a value for t, say $t = 12$ years. Be sure to tell the calculator you are solving for **A** by listing it as the variable in the solve routine.
 	
(If necessary, refer to Section 1.2.2 of this *Guide* to review the instructions for using the TI-82 solve routine.)
	`12→T`
	`                 12`
	`solve(5.833333A∕`
	`(1-.932583^T)-52`
	`0,A,50)`
	`        50.56514655`

Choose another value for t, say $t = 16$ years.
 	
Solve for A by recalling the **solve** expression with 2nd ENTER 2nd ENTER, edit the guess if necessary, and press ENTER to find $A = \$59.963$ thousand.
	`16→T`
	`                 16`
	`solve(5.833333A∕`
	`(1-.932583^T)-52`
	`0,A,63)`
	`        59.96278172`

Repeat this process for $t = 20$, $t = 24$, $t = 28$, and $t = 30$.	`          67.07112772`
	`24→T`
	`                 24`
	`solve(5.833333A∕`
	`(1-.932583^T)-52`
	`0,A,63)`
	`        72.44786154` · `          76.51480855`
	`30→T`
	`                 30`
	`solve(5.833333A∕`
	`(1-.932583^T)-52`
	`0,A,63)`
	`        78.16010439`

TI-83 Access the solver and enter $M(t, A) - 520$ after the eqn: 0 = prompt.
 	
(If necessary, refer to Section 1.2.2 of this *Guide* to review the instructions for using the TI-82 solve routine.)
	`EQUATION SOLVER`
	`eqn:0=5.833333A∕`
	`(1-.932583^T)-52`
	`0`

Press ▽ ▽ and enter a value for t, say $t = 12$ years. Be sure to tell the calculator that you are solving for **A** by having the cursor on that line when you press ALPHA ENTER (SOLVE).	`5.833333A∕(1-…=0`
	`▪A=50.565146552…`
	`  T=12`
	`  bound=(-1ᴇ99,1…`
	`▪left-rt=0`

Enter another value for *t*, say *t* =16 years.
 	
Move the cursor to the **A** line and press ALPHA
ENTER (SOLVE) to find *A* = \$59.963 thousand.
	```5.833333A/(1-…=0```
	```▪A=59.962781721…```
	```T=16```
	``` bound=(-1E99,1…```
	```▪left-rt=0```

Repeat this process for *t* = 20,
t = 24, *t* = 28, and *t* = 30.
	```5.833333A/(1-…=0```
	```▪A=72.447861537…```
	```T=24```
	``` bound=(-1E99,1…```
	```▪left-rt=0```
		```5.833333A/(1-…=0```
		```▪A=78.160104385…```
		```T=30```
		``` bound=(-1E99,1…```
		```▪left-rt=0```

Both Make a table of the values of *t* and *A* as you find them. You need to find as many points as it takes for you to see the pattern the points are indicating when you plot them on a piece of paper.

Step 3:   You need a model to draw a contour graph using your calculator. Even though there are several models that seem to fit the data points obtained in Step 2, their use would be misleading since the real best-fit model can only be determined by substituting the appropriate values in a multivariable function. The focus of this section is to use a graph to study the relationships between input variables, not to find the equation of a model for a contour curve. Thus, we always sketch the contours on paper, not with your calculator.

📖 9.3 Partial Rates of Change

When you hold all but one of the input variables constant, you are actually looking at a function of one input variable.   Therefore, all the techniques we previously discussed can be used. In particular, the calculator's numerical derivative nDeriv can be used to find partial rates of change at specific values of the varying input variable. Although your calculator does not give formulas for derivatives, you can use it as discussed in Sections 4.4.1 and/or 4.4.2 of this *Guide* to check your algebraic formula for a partial derivative.

9.3.1 NUMERICALLY CHECKING PARTIAL DERIVATIVE FORMULAS

The basic concept behind checking your partial derivative formula is that your formula and the calculator's formula computed with nDeriv should have the same outputs when each is evaluated at several different randomly-chosen inputs.

You can use the methods in Section 9.1.3 of this *Guide* to evaluate each formula at several different inputs and determine if the same numerical values are obtained from each formula. If so, your formula is *probably* correct.

You may find it more convenient to use your calculator's TABLE. If so, you must remember that when using the table, your calculator considers X as the variable that is changing. When finding a partial derivative formula, all other variables are held constant except the one that is changing. So, simply call the changing variable X and proceed as in Chapter 4.

Suppose you have a multivariable function $A(P, t) = P(1.061363551)^t$ and find, using derivative formulas, that

$$\frac{\partial A}{\partial t} = P \, (\ln 1.061363551) \, (1.061363551)^t$$

We now illustrate how to check to see if you obtained the correct formula for $\frac{\partial A}{\partial t}$.

Have $A(P, t) = P(1.061363551)^t$ in Y1. Since t is the variable that is changing, replace T with X.
 	
Enter $\frac{\partial A}{\partial t}$, as computed by the calculator, in Y2. (Notice that you must use X as the input variable.)
	`Y1▪P(1.061363551`
	`)^X`
	`Y2▪nDeriv(Y1,X,X`
	`)`
	`Y3=`
	`Y4=`
	`Y5=`
	`Y6=`

Enter your answer for $\frac{\partial A}{\partial t}$ in Y3. (Remember to use X instead of t.)
 	
Since you are checking to see if Y2 and Y3 are the same, turn Y1 off.
	`Y1=P(1.061363551`
	`)^X`
	`Y2▪nDeriv(Y1,X,X`
	`)`
	`Y3▪P(ln 1.061363`
	`551)(1.061363551`
	`)^X`
	`Y4=`

Return to the home screen. Since P is constant, store *any* reasonable value except 0 in P. (If you prefer, repeat what follows for $P = 1000$, $P = 7500$, and then $P = 190{,}000$. This, however, is not really necessary.)	`5000→P` `           5000`

Press 2nd WINDOW (TblSet) and choose ASK in the Indpnt: location.
 	
(Remember that with ASK, it does not matter what the Tblmin and ΔTbl settings are.)
	`TABLE SETUP`
	`  TblMin=1`
	`  ΔTbl=1`
	`Indpnt: Auto `**`ASK`**
	`Depend: `**`AUTO`**` Ask`

Press 2nd GRAPH (TABLE) and input some reasonable values for $t = $ X. Input at least 4 values for t. Since the values shown for Y2 and Y3 are approximately the same, our formula for $\frac{\partial A}{\partial t}$ is *probably* correct.	X Y2 Y3 2 335.44 335.44 4 377.87 377.87 8 479.51 479.51 10 540.16 540.16 25 1319.7 1319.7 `X=`

- Note that if you have rounded a model fit by the calculator before computing a partial derivative, then the Y2 and Y3 columns will very likely be *slightly* different.

- To check your formula for $\frac{\partial A}{\partial P}$, repeat the above procedure, but this time replace P by X and store any reasonable value (except 0) in T before going to the table. Remember that the variable that is changing always appears in the denominator of the partial derivative symbol.

9.3.2 VISUALIZING AND ESTIMATING PARTIAL RATES OF CHANGE

A partial rate of change of a multivariable function (evaluated at a specific point) is the slope of the line tangent to a cross-sectional model at a given location. We illustrate this concept with the two elevation cross-sectional models $E(0.8, n)$ and $E(e, 0.6)$.

Have $E(0.8, n)$ in Y1 and $E(e, 0.6)$ in Y2. (These models were found in Sections 9.1.1 and 9.1.2 of this *Guide*.)
 	
Turn Y1 on and Y2 off. (Remember that a function in the Y= list is "on" when its equals sign is darkened and "off" when its equals sign is not darkened.)
 	
Also make sure all **Stat Plots** are off.
	`Y1=-2.4999999999`
	`999X^2+2.4970588`
	`235293X+799.4897`
	`0588235`
	`Y2=-10.124103622`
	`553X^3+21.347160`
	`321612X^2+-13.97`
	`2329485096X+802.`

Press [WINDOW]. Your window should be set to values close to those shown to the right.
 	
Return to the home screen with [2nd] [MODE] (QUIT).
	`WINDOW FORMAT`
	`Xmin=-.15`
	`Xmax=1.65`
	`Xscl=1`
	`Ymin=794.99`
	`Ymax=803.51`
	`Yscl=1`

Draw the tangent line to $E(0.8, n)$ at $n = 0.6$ with the instruction **Tangent** (Y1 , 0.6). Recall that the tangent instruction is accessed with [2nd] [PRGM] (Draw) [5] (Tangent().	
You can obtain an estimate of the slope of this tangent line with the following procedure: Press [2nd] [TABLE] (CALC) [6] (dy/dx). Use ◀ or ▶ to move the cursor as close as possible to $n = x = 0.6$. (Your window will probably be set in such a way that you will not be able to locate $n = X = 0.6$ exactly.)	`X=.59680851  Y=800.08952`
Press [ENTER] to obtain an estimate of the slope of the tangent line. Remember that this is only an *estimate* of the partial rate of change $\dfrac{\partial E(0.8,n)}{\partial n}$ at $n = 0.6$.	`dy/dx=-.4869837`

Next, Turn Y1 off and turn Y2 on.
 	
Return to the home screen.
	`Y1=-2.4999999999`
	`999X^2+2.4970588`
	`235293X+799.4897`
	`0588235`
	`Y2=-10.124103622`
	`553X^3+21.347160`
	`321612X^2+-13.97`
	`2329485096X+802.`

Draw the tangent line to $E(e, 0.6)$ at $e = 0.8$ with the instruction **Tangent** (Y2 , 0.8).	

Press $\boxed{\text{2nd}}$ $\boxed{\text{TABLE}}$ (CALC) $\boxed{6}$ (dy/dx). Use $\boxed{\blacktriangleleft}$ or $\boxed{\blacktriangleright}$ to move the cursor as close as possible to $e = x = 0.8$. (Your window will probably be set in such a way that you will not be able to locate $e = X = 0.8$ exactly.)	X=.80744681 Y=800.11555
Press $\boxed{\text{ENTER}}$ to obtain an estimate of the slope of the tangent line. Remember that this is only an *estimate* of the partial rate of change $\dfrac{\partial E(e,0.6)}{\partial e}$ at $e = 0.8$.	dy/dx=.69920723

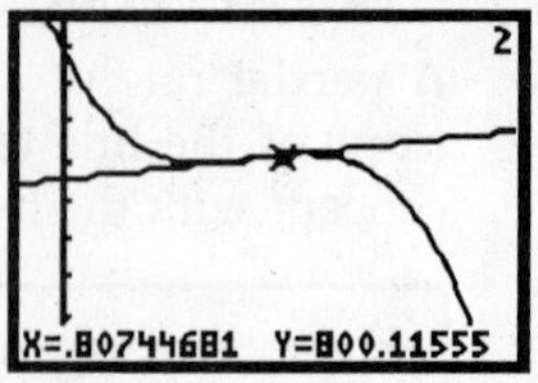

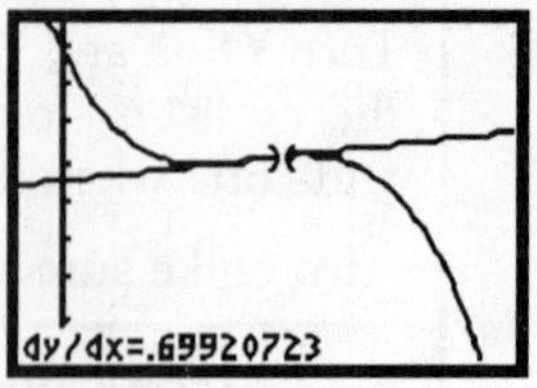

9.3.3 FINDING PARTIAL RATES OF CHANGE USING CROSS-SECTIONAL MODELS

The procedure described above gave estimates of the partial rates of change. You will obtain a more accurate answer if you use your calculator's numerical derivative on the home screen to evaluate partial rates of change at specific input values.

Have $E(0.8, n)$ in Y1 and $E(e, 0.6)$ in Y2.	Y1∎-2.4999999999 999X^2+2.4970588 235293X+799.4897 0588235 Y2∎-10.124103622 553X^3+21.347160 321612X^2+-13.97 2329485096X+802.
On the home screen, enter the instruction shown to the right to find $\dfrac{\partial E(0.8,n)}{\partial n}$ evaluated at $e = 0.8$, $n = 0.6$. (We are using X as the input variable because the cross-sectional models are in the Y= list in terms of X.)	nDeriv(Y1,X,.6) -.502941175

On the home screen, enter the second instruction shown to the right to find $\dfrac{\partial E(e,0.6)}{\partial e}$ evaluated at $e = 0.8$, $n = 0.6$.	nDeriv(Y1,X,.6)
	          -.502941175
	nDeriv(Y2,X,.8)
	          .74483795

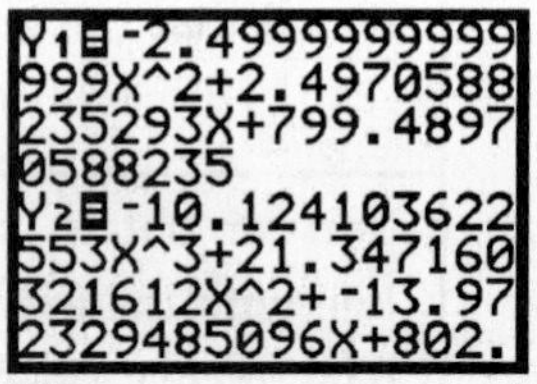

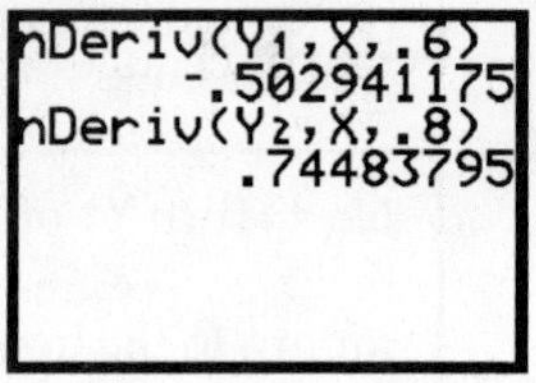

9.4 Compensating for Change

As you have just seen, your calculator finds numerical values of partial derivatives using its nDeriv function. This technique can also be very beneficial and help you eliminate many potential calculation mistakes when you find the rate of change of one input variable with respect to another input variable (that is, the slope of the tangent line) at a point on a contour curve.

9.4.1 EVALUATING PARTIAL DERIVATIVES OF MULTIVARIABLE FUNCTIONS

TIONS   The previous section of this *Guide* indicated how to estimate and evaluate partial derivatives using cross-sectional models. Your calculator can be very helpful when evaluating partial derivatives calculated directly from multivariable function formulas. The most important thing to remember is that you must supply the name of the input variable that is changing and the values at which the partial derivative is evaluated.

Consider, for instance, the formula for a person's body-mass index:

$$B(h,w) = \frac{0.45w}{0.00064516h^2}$$

where h is the person's height in inches and w is the person's weight in pounds. We first find the partial derivatives $\frac{\partial B}{\partial h}$ and $\frac{\partial B}{\partial w}$ at a specific height and weight and then use those values in the next section of this *Guide* to find the value of the derivative $\frac{dw}{dh}$ at that particular height and weight.

Suppose the person in this example is 5 feet 7 inches tall and weighs 129 pounds.

In the previous section, we had to use X as the input variable because we were finding and graphing cross-sectional models. However, here we are given the multivariable formula.
 	
Enter B(h, w) in Y1 using the same letters that appear in the formula above.
	`Y1◻.45W/(.000645`
	`16H²)`
	`Y2=`
	`Y3=`
	`Y4=`
	`Y5=`
	`Y6=`
	`Y7=`

Let's first find $\frac{\partial B}{\partial h}$ at $(h, w) = (67, 129)$. Because h is changing, w is constant. Return to the home screen and store the value of the input that is held constant.
 	
Find the value of $\frac{\partial B}{\partial h}$ at $h = 67$ and $w = 129$ by evaluating the expression shown to the right.
 	
(Store this value for later use as **N**.)
	`129→W`
	`                129`
	`nDeriv(Y1,H,67)`
	`        -.598329449`
	`Ans→N`
	`        -.598329449`
	
	H is the variable, and W is constant.

We now find $\frac{\partial B}{\partial w}$ at $(h, w) = (67, 129)$. Because w is changing, h is constant. Return to the home screen and store the value of the input that is held constant.
 	
Find the value of $\frac{\partial B}{\partial w}$ at $h = 67$ and $w = 129$ by evaluating the expression shown to the right.
 	
(Store this value for later use as **D**.)
	`67→H`
	`                 67`
	`nDeriv(Y1,W,129)`
	`          .155380128`
	`Ans→D`
	`          .155380128`
	
	W is the variable, and H is constant.

9.4.2 COMPENSATING FOR CHANGE   When one input of a two-variable multivariable function changes by a small amount, the value of the function is no longer the same as it was before the change. The methods illustrated below show how to determine the amount by which the other input must change so that the output of the function remains at the value it was before any changes were made.

We use the body-mass index formula of the previous section of this *Guide* to illustrate the procedure for calculating the compensating amount.

We first determine $\frac{dw}{dh}$ at the point (67, 129) on the contour curve corresponding to the person's current body-mass index. The formula is $\frac{dw}{dh} = \frac{-\partial B}{\partial h} \div \frac{\partial B}{\partial w}$. An easy way to remember this formula is that whatever variable is in the numerator of the derivative (in this case, w) is the same variable that appears in the denominator of the partial derivative in the denominator of the quotient. This is why we stored $\frac{\partial B}{\partial w}$ as D (for denominator). The partial derivative with respect to the other variable (here, h) appears in the numerator of the quotient. Don't forget to put a minus sign in front of the numerator.

In the previous section, we stored $\frac{\partial B}{\partial h}$ as N and $\frac{\partial B}{\partial w}$ as D. So, $\frac{dw}{dh} = \frac{-\partial B}{\partial h} \div \frac{\partial B}{\partial w} = {}^-\text{N} \div \text{D}$. The rate of change is about 3.85 pounds per inch.	```\n-N/D\n 3.850746274\n```
To estimate the weight change needed to compensate for growths of 0.5 inch, 1 inch, and 2 inches if the person's body-mass index is to remain constant, you need to find $$\Delta w \approx \frac{dw}{dh}(\Delta h)$$ at the various values of Δh.	```\n 3.850746274\n-N/D*.5\n 1.925373137\n-N/D*1\n 3.850746274\n-N/D*2\n 7.701492549\n```

---

[1]A program to graph a multivariable function with two input variables (as well as many other Texas Instruments programs, are available at http://www.ti.com/calc/docs/80xthing.htm.

Chapter 10 Analyzing Multivariable Change: Optimization

📖 10.2 Multivariable Optimization

As you might expect, multivariable optimization techniques on your calculator are very similar to those that were discussed in Chapter 5. The basic difference is that the algebra required to get the expression that comes from solving a system of equations in several unknowns down to one equation in one unknown is sometimes fairly difficult. However, once your equation is of that form, all the optimization procedures are the same as previously discussed.

10.2.1 FINDING CRITICAL POINTS
To find critical points for a multivariable function (points at which maxima, minima, or saddle points occur), find the point or points at which the partial derivatives with respect to each of the input variables are zero. Once you use derivative formulas to find the partial derivative with respect to each input variable, set these partial derivatives formulas each equal to 0, and use algebraic methods to obtain one equation in one unknown input. Then, you can use the your calculator's solve routine to obtain the solution to that equation.

We illustrate these ideas with a model for the total daily intake of organic matter required by a beef cow grazing the Northern Great Plains rangeland:

$$I(s, m) = 8.61967 - 1.244s + 0.0897s^2 - 0.20988m + 0.035947m^2 + 0.214915sm \text{ kg per day}$$

Find the two partial derivatives and set each of them equal to 0 to obtain

$$\frac{\partial I}{\partial s} = {}^-1.244 + 0.1794s + 0.214915m = 0 \text{ and } \frac{\partial I}{\partial m} = {}^-0.20988 + 0.071894m + 0.214915s = 0$$

Solve one of the equations for one of the variables, say $\frac{\partial I}{\partial s}$ for s, to obtain

$$s = \frac{1.244 - 0.214915m}{0.1794}$$

(Remember that to solve for a quantity means that it must be by itself on one side of the equation and the other side of the equation cannot contain that letter at all.) Now, let your calculator do the rest of the work.

Warning: Everything that you do in your calculator depends on the partial derivative formulas that you have found using derivative rules. Double check your analytic work before you use any of the methods illustrated below.

Enter the expression for s, using the *same* letters that are in the problem situation, in Y1. Remember that any numerator, denominator, or power that consists of more than one number or letter should be enclosed in parentheses.	```Y1◻(1.244-.21491 5M)/.1794 Y2= Y3= Y4= Y5= Y6= Y7=```
Type the expression for the *other* partial derivative, $\frac{\partial I}{\partial m}$, exactly as you have it written on your paper, in Y2. (The expression you type should be equal to zero.)	```Y1◻(1.244-.21491 5M)/.1794 Y2◻-.20988+.0718 94M+.214915S Y3= Y4= Y5= Y6=```

Now, put the cursor in the Y3 location. Type Rcl Y2 and press ENTER. [Access with 2nd STO▸ (Rcl).] Next, replace *every* S in Y3 with Y1. What you have just done is substitute the expression for *s* in the other partial derivative!	```Y1◻(1.244-.21491 5M)/.1794 Y2◻-.20988+.0718 94M+.214915S Y3◻-.20988+.0718 94M+.214915Y1 Y4= Y5=```
TI-82 The only thing that is left is to solve the equation Y3 = 0. Use the solve routine to do this. You can try several different guesses and see that they all result in the same solution. The best approach, however, is to note that Y3 is a linear equation and therefore has only one root.	```solve(Y3,M,5) 6.899884051 solve(Y3,M,35) 6.899884051 solve(Y3,M,0) 6.899884051```
TI-83 The only thing that is left is to solve the equation Y3 = 0. Use the equation solver to do this. You can try several different guesses and see that they all result in the same solution. The best approach, however, is to note that Y3 is a linear equation and therefore has only one root.	```Y3=0 ▪M=6.8998840512... bound={-1E99,1... ▪left-rt=0```
Both Return to the home screen and type M to recall the value you just found. Remember that you have the expression for *s* in Y1, so just type Y1 to find the value of *s*. The calculator, however, doesn't know this is *s*, so store the answer in memory location *s*. Note that if you now find the value of Y2, it should be 0.	``` 6.899884051 Y1 -1.331597441 Ans→S -1.331597441 Y2 0```
Either type the multivariable function $I(s, m)$ in one of the unused locations in the Y= list and evaluate it or just type $I(s, m)$ on the home screen. (You already have the values you found for *s* and *m* stored in those variables.) What all this has done is locate a critical point of $I(s, m)$ at $s \approx {}^-1.33$ and $m \approx 6.90$.	```8.61967-1.244S+. 0897S²-.20988M+. 035947M²+.214915 SM 8.723849776```

10.2.2 THE DETERMINANT TEST   Now that you have found one or more critical points, the next step is to classify those point(s) as points at which a maximum, a minimum, or a saddle point occurs. The Determinant Test will often give you the answer. Because this test uses derivatives, the calculator's numerical derivative nDeriv can help you.

For the beef cattle example in Section 10.2.1 with critical point $s \approx {}^-1.33$ and $m \approx 6.90$, you now need to calculate some second partial derivatives evaluated at this point. Recall that we have $\frac{\partial I}{\partial m} = I_m$ in Y2. However, Y1 does not contain $I_s = \frac{\partial I}{\partial s}$.

Clear Y1 and enter $\frac{\partial I}{\partial s} = {}^-1.244 + 0.1794s + 0.214915m$. (We will not use Y3 again, so you can also clear that location if you wish.)	```Y1◻-1.244+.1794S +.214915M Y2◻-.20988+.0718 94M+.214915S Y3= Y4= Y5= Y6=```

Return to the home screen, and recall **S** and **M** to be certain that the critical point values are still stored in those locations. If not, store the critical values in those locations. **This is a very important step!** (You will have better accuracy with the Determinant Test if you use several decimal places for the values.)	```S .7854191806 -1.331597→S -1.331597 6.8998841→M 6.8998841```
Now, if we take the derivative of Y1 = I_s with respect to s, we will have I_{ss}, the second partial derivative of $I(s, m)$. Find $I_{ss} \approx 0.1794$. If we take the derivative of Y2 = I_m with respect to s, we will have I_{ms}, the partial derivative of $I(s, m)$ with respect to m and then s. Find $I_{ms} \approx 0.2149$.	```nDeriv(Y1,S,-1.3 31597) .1794 nDeriv(Y2,S,-1.3 31597) .214915``` Be certain the value you type after the variable matches the input that is varying.
Now, if we take the derivative of Y2 = I_m with respect to m, we will have I_{mm}, the second partial derivative of $I(s, m)$. Find $I_{mm} \approx 0.0719$. If we take the derivative of Y1 = I_s with respect to m, we will have I_{sm}, the partial derivative of $I(s, m)$ with respect to s and then m. Find $I_{sm} \approx 0.2149$.	```nDeriv(Y2,M,6.89 98841) .071894 nDeriv(Y1,M,6.89 98841) .214915``` Be certain the value you type after the variable matches the input that is varying.
Now, find the value of $D \approx (0.1794)(0.0719) - (0.2149)^2$. Since $D < 0$, the Determinant Test tells us that the point $(s, m, I) \approx (^-1.33, 6.90, 8.72)$ a saddle point.	```.1794*.071894-.2 149152 -.0332906736```

- The values of the second partial derivatives were not very difficult to determine without the calculator in this example. However, with a more complicated function, we strongly suggest using the above methods to provide a check on your analytic work to avoid making simple mistakes.

10.3   Optimization Under Constraints

Optimization techniques on your calculator when a constraint is involved are exactly the same as those discussed in Sections 10.2.1 and Section 10.2.2 except that there is one more equation involved in the analytic process.

10.3.1 CLASSIFYING OPTIMAL POINTS UNDER CONSTRAINED OPTIMIZATION
Use the methods indicated in Section 10.2.1 to find the critical point $L = 7.35$, $K = 39.2$. We illustrate the procedure for determining if this is the point at which a maximum or minimum occurs for the Cobb-Douglas Production function $f(L, K) = 48.1L^{0.6}K^{0.4}$ subject to the constraint $g(L, K) = 8L + K = 98$.

Find $f(L, K)$ at $L = 7.35$, $K = 39.2$. We now use the methods of Section 10.2.2 to test close points that are on the constraint.	7.35→L 　　　　　7.35 39.2→K 　　　　　39.2 48.1L^.6*K^.4 　　690.6084798

We now test *close* points.   You need to remember that whatever close points you test, they must be near the critical point and *they must be on the constraint g(L,K)*. Be wary of rounding during the following procedure. Rounding of intermediate calculations and/or critical points can give a false result when the "close" point is very close to the optimal point.

Choose a new value of L that is less than L, say 7.3. Determine the value of K so that $8L + K = 98$: 　　$8(7.3) + K = 98$, so $K = 98 - 58.4 = 39.6$. Now find $f(L, K)$ at $L = 7.3$, $K = 39.6$.	7.3→L:39.6→K 　　　　　39.6 48.1L^.6*K^.4 　　690.5845641
Choose another nearby value of L, but this time choose one that is more than L, say 7.4. Determine the value of K so that $8L + K = 98$: 　　$8(7.4) + K = 98$, so $K = 98 - 59.2 = 38.8$. Now find $f(L, K)$ at $L = 7.4$, $K = 8.8$.	7.4→L:38.8→K 　　　　　38.8 48.1L^.6*K^.4 　　690.5844554

Since $f(7.35, 39.2) \approx 690.608$ is greater than both $f(7.3, 39.6) \approx 690.585$ and $f(7.4, 38.8) \approx 690.585$, we conclude that $L = 7.35$ thousand worker hours, $K = \$39.2$ thousand is the point at which the maximum value of $f(L, K)$ occurs. Rounding this answer to make sense in the context of the problem, we find that 691 mattresses is a maximum production level.

- When you test close points, it is best to use all the decimal places provided by the calculator for the optimal values when classifying optimal points since rounding will, in many cases, give inaccurate or misleading results in the classification.

Chapter 11 Dynamics of Change: Differential Equations and Proportionality

📖 11.1 Differential Equations and Accumulation Functions

Many of the differential equations we encounter have solutions that can be found by determining an antiderivative of the given rate-of-change function. So, many of the techniques we learned with the calculator's numerical integration function apply to this chapter. (See Chapters 6 and 7 of this *Guide*.)

11.1.1 EULER'S METHOD FOR $dy/dx = g(x)$

You may encounter some differential equations that cannot be solved by standard methods. You may want to draw an accumulation graph for a differential equation without first finding an antiderivative. In either of these cases, Euler's method is helpful. Euler's method relies on the use of the derivative of a function to approximate the change in the function. Recall from Section 5.3 of *Calculus Concepts* that the approximate change in *f* is the rate of change of *f* times a small change in *x*. That is,

$$f(x + h) - f(x) \approx f'(x){\cdot}h \text{ where represents the small change in } x.$$

Now, if we let $b = x + h$, and $x = a$, the above expression becomes

$$f(b) - f(a) \approx + f'(a){\cdot}(b - a) \quad \text{or} \quad f(b) \approx f(a) + (b - a){\cdot}f'(a)$$

The first values used for the coordinates of the point (a, b) will be given to you and are often called the initial condition. Then use the formula given above to involve the slope of the tangent line at *a* to approximate the change in the function between *a* and *b*. When *h*, the distance between *a* and *b*, is fairly small, Euler's method will often give close numerical estimates of points on the solution to the differential equation containing $f'(x)$.

Be wary of the fact that there is some error involved in each step of the approximation process that is compounded when each result is used to obtain the next result.

We illustrate Euler's method with the differential equation giving the total sales, in billions of dollars, of a computer product:

$$\frac{dS}{dt} = \frac{6.544}{\ln(t + 1.2)} \text{ billion dollars per year}$$

where *t* is the number of years after the product is introduced. Because Euler's method involves a repetitive process, a program that performs the calculations used to find the approximate change in the function can save you time and eliminate computational errors.

Before using this program, you must have the differential equation in location Y1 of the Y= list with X as the input variable.
 	
Note: If the differential equation is a function of two variables, those variables must be called *x* and *y* when using program EULER.
	```Y1■6.544/ln (X+1.2)```
	```Y2=```
	```Y3=```
	```Y4=```
	```Y5=```
	```Y6=```
	```Y7=```

Add program EULER to your list of programs. (The program is found in the TI-82/83 Appendix.)	```EXEC EDIT NEW```
	```1:AUTOSCL```
	```2:DIFF```
	```3:EULER```
	```4:LOGISTIC```
	```5:LSLINE```
	```6:NUMINTGL```
	```7↓SECTAN```

Run the program. Each time the program stops for input or for you to view a result, press ENTER to continue. We are told in this example to use 16 steps. Enter this value. We are also told to use steps of size 0.25. Enter this value.	``` HAVE DY/DX IN Y1 NUMBER OF STEPS= 16 STEP SIZE= .25 ```
The initial condition is given as the point (1, 52.3). Enter these values when prompted for them.	``` NUMBER OF STEPS= 16 STEP SIZE= .25 INITIAL INPUT= 1 INITIAL OUTPUT= 53.2 ```

The first application of the formula gives an us an estimate for the value of the quantity function at $x = 1.25$:
 	
$\quad S(1.25) \approx 55.275.$
	``
	INITIAL INPUT=
	1
	INITIAL OUTPUT=
	53.2
	INPUT,OUTPUT IS
	        1.25
	  55.27493782
	``

Continue pressing ENTER to obtain more estimates of points on the quantity function S. Record the input values and the output estimates on paper as they are displayed.	```
	  71.70209612
	INPUT,OUTPUT IS
	        4.75
	  72.64207417
	INPUT,OUTPUT IS
	          5
	  73.55942757
	``

When the number of steps has been completed, the program draws a graph of the points *(input, output estimate)* connected with straight line segments. This is an approximation to the graph of the solution of the differential equation.	

11.1.2 EULER'S METHOD FOR $dy/dx = h(x, y)$ Program EULER can be used when the differential equation is a function of x and y with $y = f(x)$. Follow the same process as illustrated in Section 11.1.1 of this *Guide*, but enter $\dfrac{dy}{dx}$ in Y1 in terms of both X and Y.

If the differential equation is written in terms of variables other than x and y, let the derivative symbol be your guide as to which variable is the input and which is the output. For instance, if the rate of change of a quantity is given by $\dfrac{dP}{dn} = 1.346P \cdot (1 - n^2)$, you would enter Y1= 1.346Y(1 − X)², using ALPHA 1 (Y) to type Y.

When the differential equation is given in terms of x and y, such as $\dfrac{dy}{dx} = 5.9x - 3.2y$, enter Y1= 5.9X − 3.2Y. The differential equation may be given in terms of y only. For instance, if $\dfrac{dy}{dx} = k(30 - y)$ where k is a constant, enter Y1= K(30 − Y). Of course, you need to store a value for k or substitute a value for k in the equation before using the program. It is always better to store the exact value for the constant instead of using a rounded value.

TI-82/TI-83 Calculator Appendix

Programs listed below are referenced in *Part A* of this *Guide*. They should be transferred to you via a cable using the LINK mode of a TI-82 or a TI-83, transferred to your calculator using the TI-GRAPH LINK™ cable and software for a PC or Macintosh computer and a disk containing these programs, or, as a last resort, typed in your calculator. Refer to your owner's *Guidebook* for instructions on typing in the programs or transferring them via a cable from another calculator.

(Instructors who have the TI-GRAPH LINK™ software can contact the author of this *Guide* at ibbrh@clemson.edu and request that the programs be sent to them via e-mail or on a computer disk. Be sure to specify whether you use a Macintosh or a PC-compatible computer for either method of obtaining the programs. All programs can then easily be transferred to students.)

The programs and the chapter of *Calculus Concepts* in which each program is first referenced are listed below. The programs given below run on both the TI-82 and TI-83 graphing calculators, but not all the programs are needed for the TI-83. All these programs can be transferred directly from a TI-82 to a TI-83 and vice-versa.

PROGRAM NAME	PROGRAM SIZE (bytes)	CHAPTER FIRST REFERENCED	CALCULATOR
AUTOSCL	223	1	TI-82
DIFF	445	1	TI-82, TI-83
LOGISTIC	1622	2	TI-82
NUMINTGL	991	6	TI-82, TI-83
SINREG	1091	8	TI-82
EULER	237	11	TI-82, TI-83
OPTIONAL			
LSLINE	473	1	TI-82, TI-83
SECTAN	449	4	TI-82, TI-83

The code for each of the programs follows. If you have to type in these programs rather than having them transferred from another calculator or a computer, it is strongly suggested that you compare the line-by-line instructions given in the code with what you type in your calculator. Even one misplaced symbol or letter will cause the program to not properly execute.

AUTOSCL   • Program
```
:ClrHome
:Disp "HAVE F(X) IN Y1"
:ClrList L5,L6
:Disp ""
:Input "Xmin? ",A
:A→Xmin
:Input "Xmax? ",B
:B→Xmax
:iPart ((B-A)/20)→W
:If W=0:0.1→W
:seq(X,X,A,B,W)→L5
:Y1(L5)→L6
:min(L6)→P
:max(L6)→Q
:Q-P→T
:P-.15T→Ymin
:Q+.15T→Ymax
:iPart (abs (Ymax-Ymin)/10)→R
:W→Xscl:R→Yscl
:DispGraph
```

EULER   • Program
```
:ClrHome
:ClrList L1,L2
:FnOff
:Disp "HAVE DY/DX IN Y1"
:Disp ""
:Input "NUMBER OF STEPS= ",N
:Input "STEP SIZE= ",H
:Input "INITIAL INPUT=  ",X
:Input "INITIAL OUTPUT= ",Y
:For(I,1,N,1)
:X→L1(I)
:Y→L2(I)
:Y1→T
:X+H→X
:Y+H*T→Y
:Disp "INPUT,OUTPUT IS"
:Disp X
:Disp Y
:Pause
:End
:X→L1(N+1):Y→L2(N+1)
:Plot1(xyLine,L1,L2,□)
:ZoomStat
```

DIFF   • Program
```
:ClrHome
:dim L1→N:N-1→dim L6
:For(H,1,N-1,1)
:L1(H+1)-L1(H)→L6(H)
:End
:For(H,1,N-2,1)
:If L6(H+1)≠L6(H)
:Goto 2
:End
:dim L2→M:M-1→dim L3
:For(A,1,M-1,1)
:L2(A+1)-L2(A)→L3(A)
:End
:M-2→dim L4
:For(B,1,M-2,1)
:L3(B+1)-L3(B)→L4(B)
:End
:Disp "HAVE X IN L1"
:Disp "HAVE Y IN L2-SEE"
:Disp "1ST DIFF IN L3,
:Disp "2ND DIFF IN L4,"
:M-1→dim L5
:1→E
:For(E,1,M,1)
:If L2(E)=0
:Goto 1:End
:For(E,1,M-1,1)
:(L3(E)/L2(E))*100→L5(E)
:End
:Disp "PERCENT CHANGE"
:Disp "IN L5":Stop
:Lbl 1
:ClrList L5
:Disp "PERCENT CHANGE"
:Disp "NOT CALCULATED"
:Stop
:Lbl 2
:Disp "INPUT VALUES NOT"
:Disp "EVENLY SPACED"
:Stop
```

```
LOGISTIC          • Program
:ClrHome
:Disp "DATA IN L₁,L₂"
:Disp ""
:Disp "ENTER CONTINUES"
:Pause
:ClrHome
:If dim L₁≠dim L₂:Then
:Disp "LIST LENGTHS"
:Disp " NOT EQUAL"
:Stop
:End
:dim L₁→M
:M→dim L₅
:max(L₂)*1.01→Z
:Z→C
:sum L₁→T
:ln L₂→L₆
:sum L₆→V
:0→N
:Repeat (C≤Z) or (N=3)
:ln (-1*L₂+C)→L₅
:sum L₅→U
:(-sum (L₅*L₁)+sum (L₆*L₁)+
    (U-V)/M*T)/(sum (L₁²)-T²/M)→B
:e^((U-V+B*T)/M)→A
:e^(-B*L₁)→L₄
:(1+A*L₄)⁻¹→L₅
:sum (L₂*L₅)/sum (L₅²)→C
:End
:C*L₅→L₃
:sum ((L₃-L₂)²)→E
:0→N
:√.5ᴇ-10→Y
:2*Y→R
:ClrHome
:Output(2,1,"STEP:")
:Output(6,1,"SSE:")
:While (R>Y) and (N<20)
:N+1→N
:Output(2,7,N)
:Output(6,6,E)
:{3,1}→dim [A]
:C→[A](1,1):A→[A](2,1)
:B→[A](3,1)
:Output(3,2,"WORKING...")
:e^(-B*L₁)→L₄
:(1+A*L₄)⁻¹→L₅
:C*L₅→L₃
:L₄*L₅²→L₆
:{3,1}→dim [C]:-2*sum ((L₂-
    L₃)*L₅)→[C](1,1)
:2*C*sum ((L₂-L₃)*L₆)→[C](2,1)
:-2*C*A*sum ((L₂-L₃)*L₆*L₁)→
    [C](3,1)
:{3,3}→dim [E]
:2*sum L₅²→[E](1,1)
:2*sum ((L₂-2*L₃)*L₆)→[E](1,2)
:[E](1,2)→[E](2,1)
```

(Program LOGISTIC continued)

```
:2*A*sum ((2*L₃-L₂)*L₆*L₁)→
    [E](1,3)
:[E](1,3)→[E](3,1)
:2*C*sum ((C*L₆-2*(L₂-L₃)*L₄*
    L₅)*L₆)→[E](2,2)
:2*C*sum ((-C*A*L₆+(2*A*L₅*L₄-
    1)*(L₂-L₃))*L₆*L₁)→[E](2,3)
:[E](2,3)→[E](3,2)
:2*C*A*sum ((C*A*L₆-(2*A*L₅*L₄-
    1)*(L₂-L₃))*L₆*L₁²)→[E](3,3)
:Output(3,2,"COMPUTING ")
:-1*[E]⁻¹*[C]→[D]
:2*E→F
:10→S
:-1→I
:While (F>E) and (S>0)
:I+1→I
:If I>5:Then
:0→S
:Else
:Output(4,2,"COMPUTING ")
:Output(4,12,I)
:.1*S→S
:S*[D]+[A]→[B]
:[B](1,1)→C
:[B](2,1)→A
:[B](3,1)→B
:e^(-B*L₁)→L₄
:C*(1+A*L₄)⁻¹→L₃
:sum (L₃-L₂)²→F
:End
:End
:Output(4,2,"          ")
:If S=0:Then
:Output(3,2,"STILL WORKING")
:-1*[C]→[D]
:([C]ᵀ*[C])*([C]ᵀ*[E]*[C])⁻¹→[B]
:10*[B](1,1)→S
:-1→I
:While (F>E) and (S>0)
:I+1→I
:If I>5:Then
:0→S
:Else
:Output(4,2,"COMPUTING ")
:Output(4,12,I)
:.1*S→S
:S*[D]+[A]→[B]
:[B](1,1)→C
:[B](2,1)→A
:[B](3,1)→B
:e^(-B*L₁)→L₄
:C*(1+A*L₄)⁻¹→L₃
:sum (L₃-L₂)²→F
:End
:End
:Output(4,2,"          ")
```

(Program LOGISTIC continued)

```
:If S=0:Then
:Output(3,2,"NO IMPROVEMENT")
:Else
:Output(3,2,"                ")
:End
:End
:F→E
:{0,0}→L₆
:S*[D](2,1)*[A](2,1)⁻¹→L₆(1)
:S*[D](3,1)*[A](3,1)⁻¹→L₆(2)
:abs L₆→L₆
:max(L₆)→R
:End
:{R,Y}
:"L/(1+Ae^(-BX))"→Y₄
:C→L
:ClrHome
:ClrList L₃,L₄,L₅,L₆
:Disp "MODEL"
:Disp "Y=L/(1+Ae^(-BX))"
:Output(4,2,"L=")
:Output(4,4,L)
:Output(6,2,"A=")
:Output(6,4,A)
:Output(7,2,"B=")
:Output(7,4,B)
```

LSLINE   • Program

```
:0→A:0→B:1→C
:"AX+B"→Y₁
:Ymax-Ymin→H
:.2H+Ymax→Ymax
:FnOff
:Text(0,0,"X TICK=",Xscl,"    Y
    TICK=",Yscl)
:Pause
:ClrHome
:Lbl 1
:Text(0,0,"GUESS  SLOPE,  Y-
    INTERCEPT")
:Pause
:FnOn
:Input "SLOPE=",A
:Input "Y-INTERCEPT=",B
:2-Var Stats
:Lbl 2
:0→S
:For(K,1,n)
:L₁(K)→X
:(L₂(K)-Y₁)²+S→S
:Line(L₁(K),L₂(K),X,Y₁)
:End
:Pause
:Disp ""
:Disp "SSE=",S
:Pause
:If C=2
:Goto 3
:Input "TRY AGAIN? 1Y 2N",C
:If C=1
:Goto 1
:LinReg(ax+b)
:"aX+b"→Y₂
:Disp ""
:Disp "PRESS ENTER TO"
:Disp "SEE YOUR LINE"
:Disp "AND BESTFIT LINE"
:Pause
:DispGraph
:Pause
:ClrHome
:Disp "NOW,PRESS ENTER"
:Disp "TO SEE ERRORS"
:Disp "FOR BESTFIT LINE"
:Pause
:a→A:b→B
:Goto 2
:Lbl 3
:Disp "SLOPE=",a
:Disp "Y-INTERCEPT=",b
:FnOff
```

```
NUMINTGL          • Program
:0→A:0→L:0→B
:ClrHome
:PlotsOff
:Disp "ENTER F(X) IN Y1"
:Disp ""
:Disp "CONTINUE?"
:Input "YES(1) NO(2) ",G
:If G=2:Stop
:Disp ""
:Disp "DRAW PICTURES?"
:Input "YES(1) NO(2) ",H
:ClrHome
:Input "LEFT ENDPOINT? ",A
:Input "RIGHT ENDPOINT? ",B
:If H=1:Then
:A→Xmin:B→Xmax
:ClrList L5,L6
:iPart ((B-A)/20)→W
:If W=0:0.1→W
:seq(X,X,A,B,W)→L5
:Y1(L5)→L6
:min(L6)→Ymin
:If Ymin>0:0→Ymin
:max(L6)→Ymax
:If Ymax<0:0→Ymax
:W→Xscl
:iPart (abs (Ymax-Ymin)/10)→Yscl
:ClrList L5,L6
:End
:Lbl 0
:ClrHome
:Disp "ENTER CHOICE:"
:Disp "LEFT RECT  (1)"
:Disp "RIGHT RECT (2)"
:Disp "TRAPEZOIDS (3)
:Input "MIDPT RECT (4) ",R
:Lbl 1
:ClrDraw
:Input "N? ",N
:(B-A)/N→W
:0→S:1→C
:Lbl 2
:If R=1:Goto 3
:If R=2:Goto 4
:If R=3:Goto 3
:If R=4:Goto 5
:Lbl 3
:A+(C-1)W→X
:X→J:X+W→L
:Goto 7
:Lbl 4
:A+CW→X
:X-W→J:X→L
:Goto 7
:Lbl 5
:If H≠1:Then
:If N>5:Then
```

(Program NUMINTGL continued)

```
:1→Z:W/2→H:A→X
:Lbl 8
:X+H→X:Y1+S→S
:A+ZW→X
:IS>(Z,N):Goto 8
:SW→S:Goto T
:End:End
:A+CW-W/2→X
:X-W/2→J
:X+W/2→L
:Goto 7
:A→G:G+W→G:G→V
:Lbl 9
:V→X:Y1→Y:U+W→X:4Y+2Y1+S→S
:U+2W→U
:If U<B:Goto 9
:G-W→X:Y1→E
:B→X:Y1→F
:(W/3)(S+E-F)→S
:Goto T
:Lbl 7
:Y1→K:K+S→S
:If H=1:Goto D
:Lbl I
:IS>(C,N)
:Goto 2
:If R=3:Then
:A→X:Y1→P
:B→X:Y1→Q
:S+(Q-P)/2→S
:End
:W*S→S
:Lbl T
:Disp "SUM=",S
:Pause
:ClrHome
:Lbl E
:Menu("ENTER CHOICE","CHANGE
    N",1,"CHANGE METHOD",0,
    "QUIT",F)
:Lbl F
:Stop
:Lbl D
:If R=3:Then
:Y1(L)→M
:Else:K→M
:End
:Line(J,0,J,K)
:Line(J,K,L,M)
:Line(L,M,L,0)
:If C=N:Pause
:Goto I
```

```
SECTAN              • Program
:ClrHome
:PlotsOff
:ClrDraw:2→R
:Disp ""
:Disp "HAVE F(X)IN Y1 AND"
:Disp "DRAW GRAPH OF F"
:Disp ""
:Disp "CONTINUE? "
:Input "YES(1) NO(2)",C
:If C=2:Stop
:Disp ""
:Disp "X-VALUE OF POINT"
:Input "OF TANGENCY? ",A
:Lbl 1
:Disp ""
:Disp "PRESS ENTER TO "
:Disp "SEE SECANT LINES"
:If R=1:Goto 2
:Disp "FROM THE LEFT"
:Goto 3
:Lbl 2
:ClrDraw
:Disp "FROM THE RIGHT"
:Lbl 3
:Disp "APPROACH TANGENT"
:Disp "LINE"
:Pause
:(Xmax-Xmin)/3→K
:If K>50:48→K
:For(J,1,5,1)
:A-K→X
:If R=1:A+K→X
:(Y1(X)-Y1(A))/(X-A)→M
:DrawF (M(X-A)+Y1(A))
:K/2→K
:End
:Pause
:If R=1:Goto 4
:1→R:Goto 1
:Lbl 4
:ClrHome
:Disp "PRESS ENTER TO"
:Disp "SEE TANGENT LINE"
:Pause
:ClrDraw
:Tangent(Y1,A)
```

```
SINREG              • Program
:ClrHome
:Disp "DATA IN L1,L2"
:Disp ""
:Disp "ENTER CONTINUES"
:Pause
:ClrHome
:dim L1→N
:If N≠dim L2:Then
:Disp "L1, L2 ARE NOT"
:Disp "SAME LENGTH"
:Stop
:End
:0→P
:Repeat P>0
:Input "PERIOD GUESS? ",P
:End
:Disp "MAXIMUM"
:Input "ITERATION? ",M
:max(min(M,16),1)→M
:0→A
:2π/P→B
:0→C
:median(L2)→D
:(10^-10)*sum (L2-D)²→P
:13→dim L3
:For(I,1,M)
:Fill(0,L3)
:For(K,1,N)
:{B*L1(K)}→L4
:{cos L4(1),sin L4(1)}→L4
:{L4(1),L4(2),L1(K)*(C*L4(1)-
    A*L4(2)),A*L4(1)+C*L4(2)+D-
    L2(K)} →L4
:{L4(1)²,L4(1)*L4(2),L4(1),L4(1)
    *L4(3),L4(1)*L4(4),L4(2)²,L4(
    2),L4(2)*L4(3),L4(2)*L4(4),L4
    (3),L4(4),L4(3)²,L4(3)*L4(4)}
    +L3→L3
:End
:[[L3(1),L3(2),L3(3),L3(4),L3(5)
    ][L3(2),L3(6),L3(7),L3(8),L3
    (9)][L3(3),L3(7),N,L3(10),L3(
    11)][L3(4),L3(8),L3(10),L3(12)
    ,L3(13)]]→[A]
:For(K,1,4)
:For(J,K+1,4)
:If abs [A](J,K)>abs [A](K,K)
:Then
:rowSwap([A],K,J)
:End:End
:If [A](K,K)≠0:Then
:*row(1/[A](K,K),[A],K)→[A]
:For(J,1,4)
:If J≠K
:*row+(-[A](J,K),[A],K,J)→[A]
:End:End:End
:Disp I
:If I>1 and [A](4,4)=0:Then
:M→I
```

(Program SINREG continued)
```
:Else
:A-[A](1,5)→A
:B-[A](4,5)→B
:C-[A](2,5)→C
:D-[A](3,5)→D
:L₃(5)*[A](1,5)+L₃(9)*[A](2,5)
:L₃(11)*[A](3,5)+L₃(13)*[A](4,5)
    →K
:If K<P:M→I
:End:End
:A→P
:√(P²+C²)→A
:If P<0:Then
: -.5*π-tan⁻¹ (C/P)→C
:Else
:.5*π-tan⁻¹ (C/P)→C
:End
:Disp ""
:Disp "ENTER CONTINUES"
:Pause
:"Asin (BX+C)+D"→Y₁
:0→dim L₃
:0→dim L₄
:ClrHome
:Disp "MODEL"
:Disp "Y=Asin (BX+C)+D"
:Output(4,2,"A=")
:Output(4,4,A)
:Output(5,2,"B=")
:Output(5,4,B)
:Output(6,2,"C=")
:Output(6,4,C)
:Output(7,2,"D=")
:Output(7,4,D)
```

PART B

TEXAS INSTRUMENTS TI-85/TI-86 GRAPHING CALCULATORS

Setup

When using this *Guide*, you should always, unless instructed otherwise, use the calculator setup specified below for both the TI-85 and TI-86. Before you begin, check the basic setup with 2nd MORE (MODE). Use the following instructions to choose the settings shown in Figure 1. Check the window format with GRAPH MORE F3 (FORMT). Choose the settings shown in Figure 2.

- If you do not have the darkened choices shown in Figures 1 and 2 (below), use the arrow keys to move the blinking cursor over the setting you want to choose and press ENTER .

- Press EXIT or 2nd EXIT (QUIT) to return to the home screen.

TI-85, 86 Basic Setup

FIGURE 1

TI-85, 86 Window Setup

FIGURE 2

For the TI-85 Return to the home screen with EXIT or 2nd EXIT (QUIT). Specify the statistical setup as shown in Figure 3 by pressing 2nd − (LIST) F1 ({) 0 F2 (}) STO▸ 7 (L) ALPHA 1 ENTER 2nd ENTER (ENTRY) ◄ 2 ENTER . (See Figure 3.) Press EXIT or 2nd EXIT (QUIT) to return to the home screen.

For the TI-86 Return to the home screen with EXIT or 2nd EXIT (QUIT). Specify the statistical setup as shown in Figure 6 by pressing 2nd − (LIST) F5 (OPS) MORE MORE MORE F3 (SetLE) ALPHA 7 (L) 1 , ALPHA 7 (L) 2 , ALPHA 7 (L) 3 , ALPHA 7 (L) 4 , ALPHA 7 (L) 5 ENTER . (See Figure 4.) Press EXIT or 2nd EXIT (QUIT) to return to the home screen.

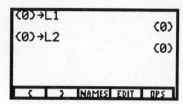

TI-85 Statistical Setup
FIGURE 3

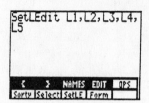

TI- 86 Statistical Setup
FIGURE 4

TI-86 Note: Because appropriate TI-86 computer linking software had not yet been perfected at the time this material was written, TI-86 screens in this *Guide* appear smaller than the TI-85 screens. Screens that appear the same on both calculators are shown for the TI-85.

Basic Operation

You should be familiar with the basic operation of your calculator. With calculator in hand, go through each of the following.

1. CALCULATING You can type in lengthy expressions; just make sure that you use parentheses when you are not sure of the calculator's order of operations. As a general rule, numerators and denominators of fractions and powers consisting of more than one term should be enclosed in parentheses.

Evaluate $\dfrac{1}{4*15+\dfrac{895}{7}}$. Evaluate $\dfrac{(-3)^4-5}{8+1.456}$.
 	
 (Use $\boxed{(-)}$ for the negative symbol and $\boxed{-}$ for the subtraction sign.)
	`1/(4*15+895/7)`
	 `        .005323193916`
	 `((-3)^4-5)/(8+1.456)`
	 `         8.0372250423`

Evaluate $e^3*0.027$ and $e^{3*0.027}$.
 	
 The calculator will assume you mean the first expression unless you use parentheses around the two values in the exponent. (It is not necessary to type in the 0 before the decimal point.)
	`e^3*.027`
	 `        .542309496926`
	 `e^(3*.027)`
	 `         1.08437089657`

2. USING THE *ANS* MEMORY Instead of again typing an expression that was evaluated immediately prior, use the answer memory by pressing $\boxed{\text{2nd}}$ $\boxed{(-)}$ (ANS).

Calculate $\left(\dfrac{1}{4*15+\dfrac{895}{7}}\right)^{-1}$ using this nice shortcut.
 	
 (If you wish to clear the home screen, press $\boxed{\text{CLEAR}}$.)
	`895/7`
	 `        127.857142857`
	 `1/(4*15+Ans)`
	 `         .005323193916`
	 `Ans⁻¹`
	 `         187.857142857`

3. ANSWER DISPLAY   When the denominator of a fraction has no more than three digits, your calculator can provide the answer in fraction form. When an answer is very large or very small, the calculator displays the result in scientific notation.

The "to a fraction" key is obtained by pressing [2nd] [X] (MATH) [F5] (MISC) [MORE] [F1] (▸Frac).	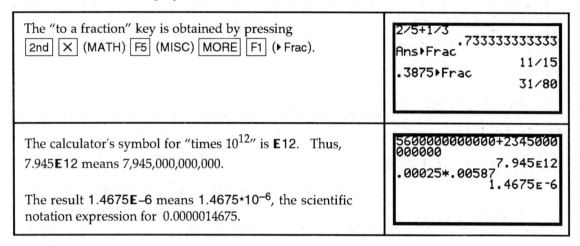
The calculator's symbol for "times 10^{12}" is **E**12. Thus, 7.945**E**12 means 7,945,000,000,000. The result 1.4675**E**–6 means $1.4675*10^{-6}$, the scientific notation expression for 0.0000014675.	

4. STORING VALUES   Sometimes it is beneficial to store numbers or expressions for later recall. To store a number, type the number on the display and press [STO▸]. (Note that the cursor automatically changes to alphabetic mode when you press [STO▸].) Next, press the key corresponding to the letter in which you wish to store the value, and then press [ENTER]. To join several short commands together, use [2nd] [.] (:).

Store 5 in A and 3 in B, and then calculate $4A - 2B$. To recall a value stored in a variable, use [ALPHA] to type the letter in which the expression or value is stored and then press [ENTER]. The value stays stored until you change it.	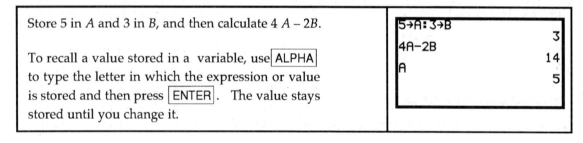

5. ERROR MESSAGES   When your input is incorrect, an error message is displayed.

If you have more than one command on a line without the commands separated by a colon (:), an error message results when you press [ENTER].	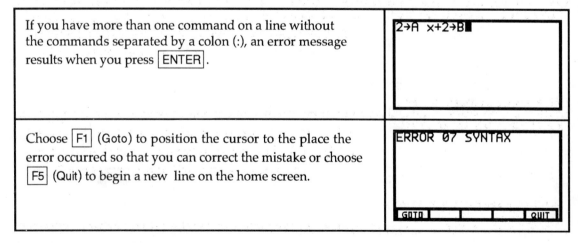
Choose [F1] (Goto) to position the cursor to the place the error occurred so that you can correct the mistake or choose [F5] (Quit) to begin a new line on the home screen.	

Chapter 1 Ingredients of Change: Functions and Linear Models

📖 1.1 Fundamentals of Modeling

There are many uses for a function that is entered in the graphing list. Graphing the function in an appropriate viewing window is one of these. Because you must enter all functions on one line (that is, you cannot write fractions and exponents the same way you do on paper) it is very important to have a good understanding of the calculator's order of operations and to use parentheses whenever they are needed.

1.1.1 ENTERING AN EQUATION IN THE GRAPHING LIST   Press $\boxed{\text{GRAPH}}$ $\boxed{\text{F1}}$

(y(x)=) to access the graphing list. Up to 99 equations can be entered in the graphing list, and the output variables are called by the names y1, y2, etc. When you intend to graph an equation you enter in the list, you must use x as the input variable.

If there are any previously entered equations that you will no longer use, clear them out of the graphing list.	Position the cursor on the line containing the equation and press $\boxed{\text{CLEAR}}$ or $\boxed{\text{F4}}$ (DELf).

Suppose you want to graph $A = 1000(1 + 0.05)^t$. Because we intend to graph this equation, the input must be called x. Type x by pressing $\boxed{\text{x-VAR}}$ or $\boxed{\text{F1}}$ (x), not the times sign $\boxed{\times}$.

TI 85 For convenience, we use the first, or y1, location in the graphing list. Enter the right hand side as $\qquad 1000(1 + 0.05)^{\wedge}x$	y1∎1000(1+.05)^x y(x)= RANGE ZOOM TRACE GRAPH x y INSf DELf SELCT▸
TI 86 For convenience, we use the first, or y1, location in the graphing list. Enter the right hand side as $\qquad 1000(1 + 0.05)^{\wedge}x$ Plot1, Plot2, and Plot3 at the top of the y(x)= list should not be darkened. If any of them are, use $\boxed{\blacktriangle}$ until you are on the darkened plot name. Press $\boxed{\text{ENTER}}$ to make the name(s) not dark.	Plot1 Plot2 Plot3 \y1∎1000(1+.05)^x y(x)= WIND ZOOM TRACE GRAPH x y INSf DELf SELCT▸

1.1.2 DRAWING A GRAPH   If you have not already done so, enter the equation in the y(x)= list using x as the input variable before drawing a graph. We now draw the graph of $y = 1000(1 + 0.05)^x$.

Remove the lower menu with $\boxed{\text{EXIT}}$, and press $\boxed{\text{F3}}$ (ZOOM) $\boxed{\text{MORE}}$ $\boxed{\text{F4}}$ (ZDECM). Notice that the graphics screen is blank.	y1∎1000(1+.05)^x y(x)= RANGE ZOOM TRACE GRAPH ZFIT ZSQR ZTRIG ZDECM ZRCL▸

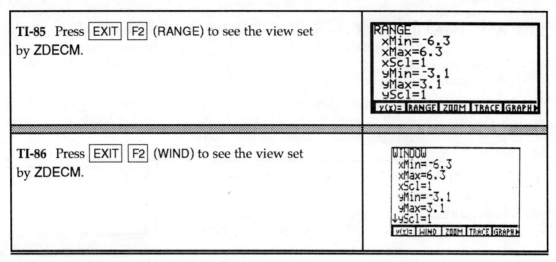

TI-85 Press [EXIT] [F2] (RANGE) to see the view set by ZDECM.

TI-86 Press [EXIT] [F2] (WIND) to see the view set by ZDECM.

- **Both** xMin and xMax are the settings of the left and right edges of the viewing screen, and yMin and yMax are the settings for the lower and upper edges of the viewing screen. xScl and yScl set the spacing between the tick marks on the x- and y-axes. The view you see is $-6.3 \le x \le 6.3$, $-3.1 \le y \le 3.1$.

Follow the procedures shown in either 1.1.3 or 1.1.4 to draw a graph with your calculator. Whenever you draw a graph, you have the option of manually changing the view or having the calculator automatically find a view of the graph.

1.1.3 MANUALLY CHANGING THE VIEW OF A GRAPH   If you do not have a good view of the graph or if you do not see the graph, change the view with one of the ZOOM options or manually set the view. (We later discuss the ZOOM options.)

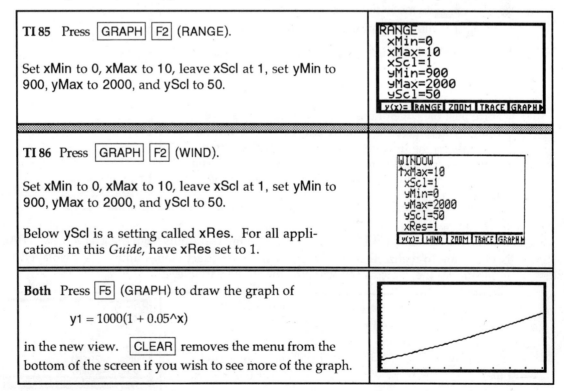

TI 85 Press [GRAPH] [F2] (RANGE).

Set xMin to 0, xMax to 10, leave xScl at 1, set yMin to 900, yMax to 2000, and yScl to 50.

TI 86 Press [GRAPH] [F2] (WIND).

Set xMin to 0, xMax to 10, leave xScl at 1, set yMin to 900, yMax to 2000, and yScl to 50.

Below yScl is a setting called xRes. For all applications in this *Guide*, have xRes set to 1.

Both Press [F5] (GRAPH) to draw the graph of

$$y1 = 1000(1 + 0.05^{\wedge}x)$$

in the new view. [CLEAR] removes the menu from the bottom of the screen if you wish to see more of the graph.

1.1.4 AUTOMATICALLY CHANGING THE VIEW OF THE GRAPH If your view of the graph is not good or if you do not see the graph, change the view using the built-in autoscaling feature of your calculator. This option will automatically find a view to see all the functions that you have turned on in the graphing list.

Be sure the function you are graphing, $y = 1000(1 + 0.05)^x$, is entered in the y1 location of the y(x)= list. (Delete all other functions that may be entered in other locations.) Before doing what follows, access the graphics menu with GRAPH.

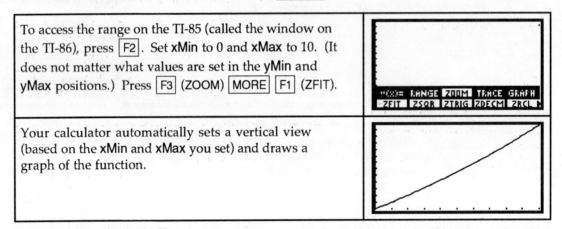

To access the range on the TI-85 (called the window on the TI-86), press F2. Set **xMin** to 0 and **xMax** to 10. (It does not matter what values are set in the **yMin** and **yMax** positions.) Press F3 (ZOOM) MORE F1 (ZFIT).

Your calculator automatically sets a vertical view (based on the **xMin** and **xMax** you set) and draws a graph of the function.

1.1.5 TRACING You can display the coordinates of certain points on the graph by tracing. The x-values shown when you trace depend on the horizontal view that you choose, and the y-values are calculated by substituting the x-values into the equation that is being graphed.

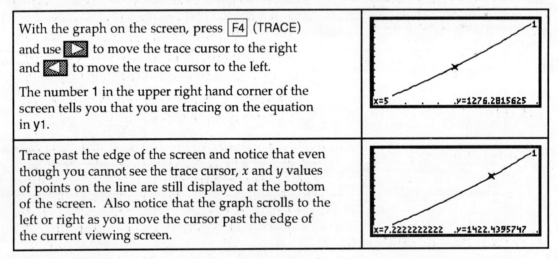

With the graph on the screen, press F4 (TRACE) and use ▶ to move the trace cursor to the right and ◀ to move the trace cursor to the left.

The number 1 in the upper right hand corner of the screen tells you that you are tracing on the equation in **y1**.

Trace past the edge of the screen and notice that even though you cannot see the trace cursor, x and y values of points on the line are still displayed at the bottom of the screen. Also notice that the graph scrolls to the left or right as you move the cursor past the edge of the current viewing screen.

1.1.6 ESTIMATING OUTPUTS You can estimate outputs from the graph using TRACE. It is important to realize that such outputs are *never* exact values unless the displayed x-value is *identically* the same as the value of the input variable.

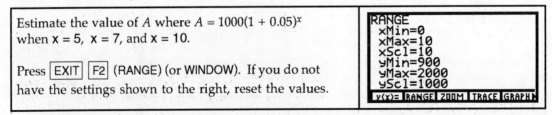

Estimate the value of A where $A = 1000(1 + 0.05)^x$ when x = 5, x = 7, and x = 10.

Press EXIT F2 (RANGE) (or WINDOW). If you do not have the settings shown to the right, reset the values.

Press F3 (ZOOM) F3 (ZOUT). After the graph finishes drawing, press ENTER to once more enlarge your view of the graph. (Press EXIT F2 (RANGE) and observe the values now defining the graphics screen.)	x=5 y=1450 Note that (5, 1450) is a point on the screen, *not* a point on the graph of the function.
Press F4 (TRACE) and use ▶ to move as close as you can to $x = 5$. (Your screen may look slightly different than the one shown to the right.) Continue pressing ▶ and notice that the values 7 and 10 cannot be obtained by tracing in this view. Therefore, choose values close to these numbers to obtain *estimates* such as A is approximately \$1386.80 when $x = 7$ and A is about \$1637.37 when $x = 10$.	x=5 y=1276.2815625 Because the number 5, *not* a value close to 5 is shown, A = \$1276.28 when $x = 5$.

- If you had used the original range (window) with **xMax** =10, and traced, you should obtain the *exact* value A = 1500 when $x = 10$ because 10, not a value close to 10, is shown when tracing.

- If you want "nice, friendly" values displayed for x when tracing, set **xMin** and **xMax** so that **xMax**–**xMin** is a multiple of 12.6, the width of the ZDECM viewing screen. For instance, if you set **xMin** = 0 and **xMax** = 12.6 in the example above, the *exact* values when **x** = 5, **x** = 7, and **x** = 10 are displayed when you trace. Another view that gives friendly values is **xMin** = ⁻5 and **xMax** = 20.2 since 25.2 = 2(12.6). Try it!

1.1.7 EVALUATING OUTPUTS   The values obtained by this evaluation process are *actual* output values of the equation, not estimated values such as those generally obtained by tracing.

Begin by entering the equation whose output you want to evaluate in the **y(x)=** list. Even though you can use any of the locations, let us say for this illustration you have **y1** = 1000(1 + 0.05^*x*).

To evaluate an output while the graph is on the screen:

Draw the graph of **y1** in a view containing the input you intend to use. (If the input value is not between **xMin** and **xMax** , an error message results when you evaluate.) Press MORE MORE F1 (EVAL), and type the input value.	Eval x=5 EVAL STPIC RCPIC
Press ENTER to display the output and see the point marked on the graph of the function. Repeat this process to find the outputs when $x = 7$ and $x = 10$.	x=5 . . . . y=1276.2815625

To evaluate an output from the home screen:

First, press EXIT until you are on the home screen. Store the input value, say 5, in x by with 5 STO▸ x-VAR ENTER. Next, type the location in the y(x)= list in which the function is stored with 2nd ALPHA 0 (y) 1. Note that the second function key is pressed before typing the letter y in order to type lower-case y.	```5→x 5``` ```y1 1276.2815625```
Warning: The TI-85 and the TI-86 distinguish between lower-case and upper-case variable names. Locations in which equations are stored in the graphing list are always referred to with *lower-case* letters.	

If you had not pressed 2nd before ALPHA and therefore entered Y1, you would be using another variable stored in the calculator's memory. (Your values may be different from the ones shown to the right or you may get an UNDEFINED error.)
 	
Note that on the TI-85, y1(5) *multiplies* y1 by 5.
	```Y1 .53```
	```y1 1000```
	```y1(5) 5000```

To evaluate another output, simply store the input value in x, type y1 and press ENTER.
 	
Evaluate y1 = 1000(1 + 0.05^x) at x = 7 and x = 10.
	``` 7```
	```y1 1407.10042266```
	```10→x 10```
	```y1 1628.89462678```

- The values obtained by either of these evaluation processes are *actual* output values of the equation, not *estimated* values such as those generally obtained by tracing.

TI-86: Evaluating outputs from the home screen   While the evaluation process indicated above works on the TI-86, advances in the TI-86 technology give you a way to evaluate functions from the home screen using fewer keystrokes than for the TI-85.

Type 2nd ALPHA 0 (y)1 and then press (5). Press ENTER. On the TI-86, typing a value in parentheses following a function evaluates the function at that input value.	```y1(5) 1276.2815625```

Evaluate y1 at $x = 7$ by recalling the previous entry with 2nd ENTER (ENTRY), edit the 5 to 7 by pressing ◀ ◀ and typing over the 5, and press ENTER.
 	
Repeat the process to evaluate y1 at $x = 10$.
	```y1(5) 1276.2815625```
	```y1(7) 1407.10042266```
	```y1(10) 1628.89462678```

📖 1.2 Functions and Graphs

When you are asked to *estimate* or *approximate* an output or an input value, you can use your calculator in the following ways:

- tracing a graph   (Sections 1.1.5, 1.1.6)
- close values obtained from a table of function values   (End of Section 1.2.2)

When you are asked to *find* or *determine* an output or an input value, you should use your calculator in the following ways:

- evaluating an output on the home screen   (Section 1.1.7)
- find a value using the **AUTO** or **ASK** features of the table   (Section 1.2.1)
- determine an input using the solver   (Section 1.2.2)

1.2.1 DETERMINING OUTPUTS

Function outputs can be determined by evaluating on the home screen, as discussed in 1.1.7. You can also evaluate functions using the calculator's **TABLE**. (On the TI-85, the table is accessed with program **TABLE**.) When you use the table, you can ask for specific output values corresponding to the inputs you enter or generate a list of input values that begin with **TblMin** and differ by **ΔTbl** and their corresponding outputs.

Let's use the **TABLE** to determine the output of the function $v(t) = 3.622(1.093)^t$ when $t = 85$. Press $\boxed{\text{GRAPH}}$ $\boxed{\text{F1}}$ (y(x)=), clear any functions, and enter 3.622(1 .093)^X in location y1. Exit.

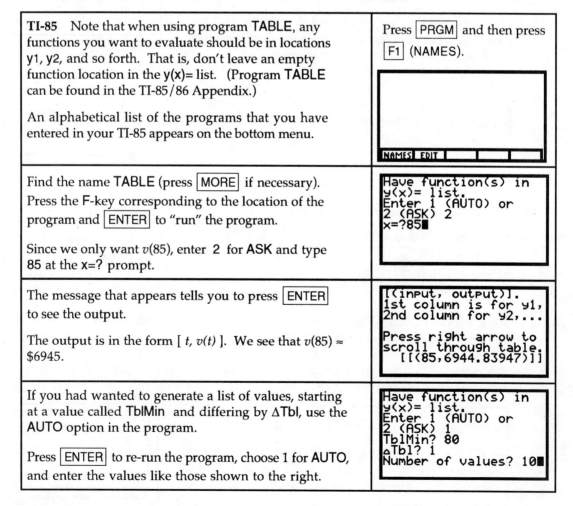

TI-85 Note that when using program **TABLE**, any functions you want to evaluate should be in locations y1, y2, and so forth. That is, don't leave an empty function location in the y(x)= list. (Program **TABLE** can be found in the TI-85/86 Appendix.) An alphabetical list of the programs that you have entered in your TI-85 appears on the bottom menu.	Press $\boxed{\text{PRGM}}$ and then press $\boxed{\text{F1}}$ (NAMES). `NAMES EDIT`

Find the name **TABLE** (press $\boxed{\text{MORE}}$ if necessary). Press the F-key corresponding to the location of the program and $\boxed{\text{ENTER}}$ to "run" the program.
 	
Since we only want $v(85)$, enter 2 for **ASK** and type 85 at the x=? prompt.
	`Have function(s) in`
	`y(x)= list.`
	`Enter 1 (AUTO) or`
	`2 (ASK) 2`
	`x=?85`

The message that appears tells you to press $\boxed{\text{ENTER}}$ to see the output.
 	
The output is in the form [t, $v(t)$]. We see that $v(85) \approx$ \$6945.
	`[(input, output)].`
	`1st column is for y1,`
	`2nd column for y2,...`
	
	`Press right arrow to`
	`scroll through table.`
	`  [[(85,6944.83947)]]`

If you had wanted to generate a list of values, starting at a value called **TblMin** and differing by **ΔTbl**, use the **AUTO** option in the program.
 	
Press $\boxed{\text{ENTER}}$ to re-run the program, choose 1 for **AUTO**, and enter the values like those shown to the right.
	`Have function(s) in`
	`y(x)= list.`
	`Enter 1 (AUTO) or`
	`2 (ASK) 1`
	`TblMin? 80`
	`ΔTbl? 1`
	`Number of values? 10`

Pressing ENTER and ▼ allows you to see the outputs that have been generated and their corresponding inputs. (If you had evaluated several different functions, their outputs appear in the same order (horizontally) as the functions in the graphing list.)	`[(82,5318.65384)]↑` `[(83,5813.28865)]` `[(84,6353.9245) ]` `[(85,6944.83947)]` `[(86,7590.70954)]` `[(87,8296.64553)]` `[(88,9068.23357)]↓`
TI-86 After entering the function $v(t)$ in y1, choose the **TABLE SETUP** menu.	Press TABLE F2 (TBLST).
To generate a list of values beginning with 80 with the table values differing by 1, enter 80 in the TblStart location, 1 in the ΔTbl location, and choose **AUTO** in the Indpnt: location. Remember that you "choose" a particular setting by positioning the blinking cursor over that setting and pressing ENTER .	`TABLE SETUP` `TblStart=80` `ΔTbl=1` `Indpnt: Auto Ask` `TABLE`
Press F1 (TABLE), and observe the list of input and output values. Notice that you can scroll through the table with ▼ , ▶ , ▲ , and/or ◀ . The table values may be rounded in the table display. You can see more of the output by moving to the value and looking at the bottom of the screen.	`x    y1` `80   4452.064` `81   4866.106` `82   5318.654` `83   5813.285` `84   6353.924` `85   6944.839` `y1=6944.8394737144` `TBLST SELCT  x    y`
Return to the **TABLE SETUP** menu with F1 (TBLST). (TblSet). To compute specific outputs rather than a list of values, choose **ASK** in the Indpnt: location. (When using **ASK**, the settings for TblMin and ΔTbl do not matter.)	`TABLE SETUP` `TblStart=80` `ΔTbl=1` `Indpnt: Auto ASK` `TABLE`
Press F1 (TABLE), type in the x-value(s) at which the function is to be evaluated, and press ENTER . You can scroll through the table with ▼ , ▶ , ▲ , and/or ◀ . Unwanted input entries can be cleared with DEL .	`x    y1` `85   6944.839` `y1=6944.8394737144` `TBLST SELCT  x    y` $v(85) \approx \$6945$

1.2.2 SOLVING FOR INPUT VALUES   Your calculator will solve for input values of an equation that you enter in the SOLVER. You can use any letter you wish for the input variable when using the SOLVER. (Of course, if you are using a function that is entered in the graphing list, the input variable is x.) You can even enter an equation consisting of several variables! You should find your calculator's solver a very valuable tool.

Press 2nd GRAPH (SOLVER) to access the SOLVER. If there is already an equation in the solver, clear it with CLEAR . You may need to press ▲ to get to the screen shown on the right before you can clear the old expression.	`eqn:` `y1   y2`

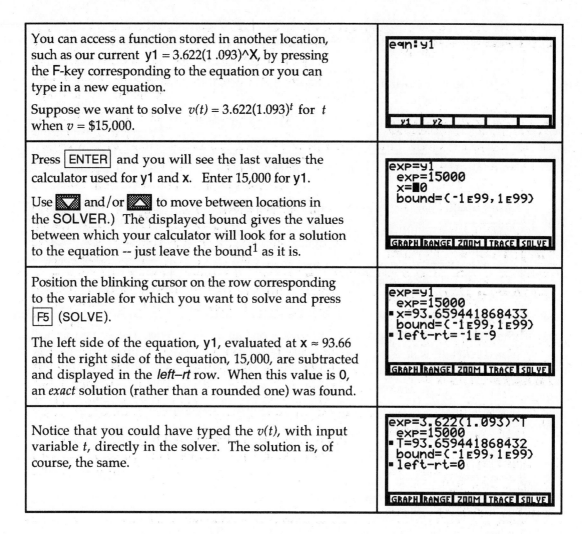

You can access a function stored in another location, such as our current $y1 = 3.622(1.093)\verb|^|X$, by pressing the F-key corresponding to the equation or you can type in a new equation.

Suppose we want to solve $v(t) = 3.622(1.093)^t$ for t when $v = \$15,000$.

Press ENTER and you will see the last values the calculator used for $y1$ and x.   Enter 15,000 for $y1$.

Use [▽] and/or [△] to move between locations in the SOLVER.) The displayed bound gives the values between which your calculator will look for a solution to the equation -- just leave the bound[1] as it is.

Position the blinking cursor on the row corresponding to the variable for which you want to solve and press [F5] (SOLVE).

The left side of the equation, $y1$, evaluated at $x \approx 93.66$ and the right side of the equation, 15,000, are subtracted and displayed in the *left–rt* row. When this value is 0, an *exact* solution (rather than a rounded one) was found.

Notice that you could have typed the $v(t)$, with input variable t, directly in the solver. The solution is, of course, the same.

If there is more than one solution to an equation, you need to give the solver an approximate location for each answer. Suppose you are given $q(x) = 8x^2 + 54.65x - 163$ and asked to find what input(s) correspond to an output of $q(x) = 154$. (The procedure outlined below also applies to finding where two functions are equal.)

Enter $8x^2 + 54.65x - 163$ in one location, say $y1$, and 154 in another location, say $y2$, in the $y(x)=$ list.

(Remember that if the input variable in the equation is not x, you must rewrite the equation in terms of x to graph using the $y(x)=$ list.)

To better obtain a guess as to where $y1$ equals (intersects) $y2$, graph the equations.

If you are not told where you want to view the graph, begin by pressing [GRAPH] [F3] (ZOOM) [F4] (ZSTD) or

---

[1] It is possible to change the bound and/or the tolerance if the calculator has trouble finding a solution to a particular equation. This, however, should not usually happen. Refer to your *Owner's Manual* for details.

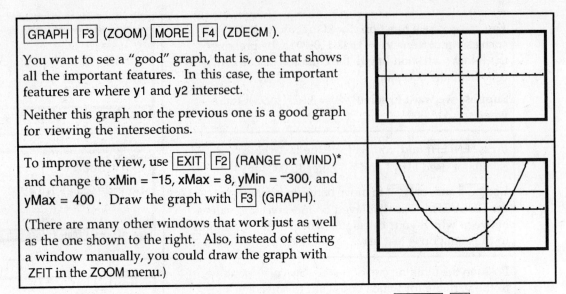

GRAPH F3 (ZOOM) MORE F4 (ZDECM).

You want to see a "good" graph, that is, one that shows all the important features. In this case, the important features are where y1 and y2 intersect.

Neither this graph nor the previous one is a good graph for viewing the intersections.

To improve the view, use EXIT F2 (RANGE or WIND)* and change to xMin = ‾15, xMax = 8, yMin = ‾300, and yMax = 400 . Draw the graph with F3 (GRAPH).

(There are many other windows that work just as well as the one shown to the right. Also, instead of setting a window manually, you could draw the graph with ZFIT in the ZOOM menu.)

*The current settings for the graphics screen are obtained with GRAPH F2 (RANGE) on the TI-85 and with GRAPH F2 (WIND) on the TI-86. When the screen appears, the word RANGE is at the top of the TI-85 screen and the word WINDOW is at the top of the TI-86 screen. Because the only difference in the calculators in this situation is the name that is used, we from this point forward refer to this key for both calculators as F2 (RANGE or WIND). In such cases, the screen shown in the right column will be the one from the TI-85.

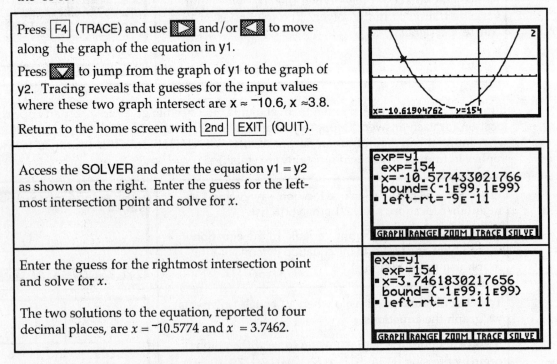

Press F4 (TRACE) and use ▶ and/or ◀ to move along the graph of the equation in y1.

Press ▼ to jump from the graph of y1 to the graph of y2. Tracing reveals that guesses for the input values where these two graph intersect are x ≈ ‾10.6, x ≈3.8.

Return to the home screen with 2nd EXIT (QUIT).

Access the SOLVER and enter the equation y1 = y2 as shown on the right. Enter the guess for the left-most intersection point and solve for x.

Enter the guess for the rightmost intersection point and solve for x.

The two solutions to the equation, reported to four decimal places, are x = ‾10.5774 and x = 3.7462.

1.2.3 GRAPHICALLY FINDING INTERCEPTS   Finding where a function graph crosses the vertical and horizontal axis can be done graphically as well as by the methods indicated in 1.2.2 of this *Guide*. Remember the process by which we find intercepts:

- To find the y-intercept of a function $y = f(x)$, set $x=0$ and solve the resulting equation.

- To find the *x*-intercept of a function $y = f(x)$, set $y=0$ and solve the resulting equation.

Also remember that an *x*-intercept of a function $y = f(x)$ has the same value as the root or solution of the equation $f(x) = 0$.

Clear all locations in the *y(x)=* list and enter in y1 $f(x) = 4x - x^2 - 2$. Draw a graph with F3 (ZOOM) MORE F4 (ZDECM). Press F2 (RANGE or WIND) and reset yMin to ‾6 for a good view of all intercepts. Press F5 (GRAPH).	
Even though it is very easy to find $f(0) = ‾2$, you can have the calculator find the *y*-intercept while viewing the graph by pressing **TI-85** MORE F1 (MATH) MORE F4 (YICPT) **TI-86** MORE F1 (MATH) MORE F2 (YICPT) **Both** View the *y*-intercept $f(0) = ‾2$.	
To graphically find an *x*-intercept, i.e., a value of *x* at which the graph crosses the horizontal axis, first press EXIT to return the menu to the bottom of the screen. **TI-85** Press F1 (MATH) F3 (ROOT) **TI-86** Press F1 (MATH) F1 (ROOT) **Both** Use ▶ to move the trace cursor near the first *x*-intercept, and press ENTER.	
To find the second *x*-intercept, press EXIT and the F-key under ROOT, use ▶ to move the trace cursor near the other *x*-intercept, and press ENTER.	

📖 1.3 Constructed Functions

Your calculator can find output values of and graph combinations of functions in the same way that you do these things for a single function. The only additional information you need is how to enter constructed functions in the graphing list. Suppose that a function $f(x)$ has been entered in y1 and a function $g(x)$ has been entered in y2.

- Enter **y1 + y2** in y3 to obtain the sum function $(f+g)(x) = f(x) + g(x)$.

- Enter **y1 – y2** in y4 to obtain the sum function $(f-g)(x) = f(x) - g(x)$.

- Enter **y1*y2** in y5 to obtain the product function $(f \cdot g)(x) = f(x) * g(x)$.

- Enter **y1 / y2** in y6 to obtain the quotient function $(f \div g)(x) = \dfrac{f(x)}{g(x)}$.

On the TI-85,

- The TI-85 interprets y1(y2) as the *product* of y1 and y2. To have the calculator graph the composite function $f(g(x))$, you should substitute $g(x)$ everywhere an x appears in $f(x)$. For instance, if $f(x) = x^2 - 1$ is in y1 and $g(x) = 0.3x + 5$ is in y2, enter y7 = y2^2 – 1 for the composite function $(f \circ g)(x)$.

On the TI-86,

- Enter y1(y2) in y7 to obtain the composite function $(f \circ g)(x) = f(g(x))$.

Both Your calculator will evaluate and graph these constructed functions. Although it will not give you an algebraic formula for a constructed function, you can check your algebra by evaluating the calculator-constructed function and your constructed function at several different points. (You will very likely have to reset the horizontal and vertical views when graphing constructed functions.)

1.3.1 GRAPHING PIECEWISE CONTINUOUS FUNCTIONS
Piecewise continuous functions are used throughout the text. It is often helpful to use your calculator to graph and evaluate outputs of piecewise continuous functions. Consider the following example.

The population of West Virginia from 1985 through 1993 can be modeled by

$$P(t) = \begin{cases} -23.514t + 3903.667 \text{ thousand people when } 85 \leq t < 90 \\ 9.1t + 972.6 \text{ thousand people when } 90 \leq t \leq 93 \end{cases}$$

where t is the number of years since 1900.

Enter the function $P(t)$, using x as the input variable, in the y1 location of the y(x)= list using the keystrokes ((–) 23.514 x-VAR + 3903.667) (x-VAR 2nd 2 (TEST) F2 (<) 90) + ((9.1 x-VAR + 972.6) ((x-VAR F5 (≥) 90) .	y1☰...9.1x+972.6)(x≥90) y(x)= RANGE ZOOM TRACE GRAPH x y INSf DELf SELCT▸ Each piece of the function and its corresponding input must be enclosed in parentheses.
Notice that the function is defined only when the input is between 85 and 93. You could find $P(85)$ and $P(93)$ to help you set the vertical view. However, we choose to let the calculator set the vertical view. Set the horizontal view xMin = 85 and xMax = 93. Press GRAPH F3 MORE F1 ZOOM (ZFIT) to graph $P(x)$.	
If you wish to see the "break" in the function where the two pieces join, the width of the screen must be a multiple of 12.6 and include 90. Since 90 – 0.5(12.6) = 83.7 and 90 + 0.5(12.6) = 96.3, change xMin and xMax to these values. Press F5 (GRAPH).	

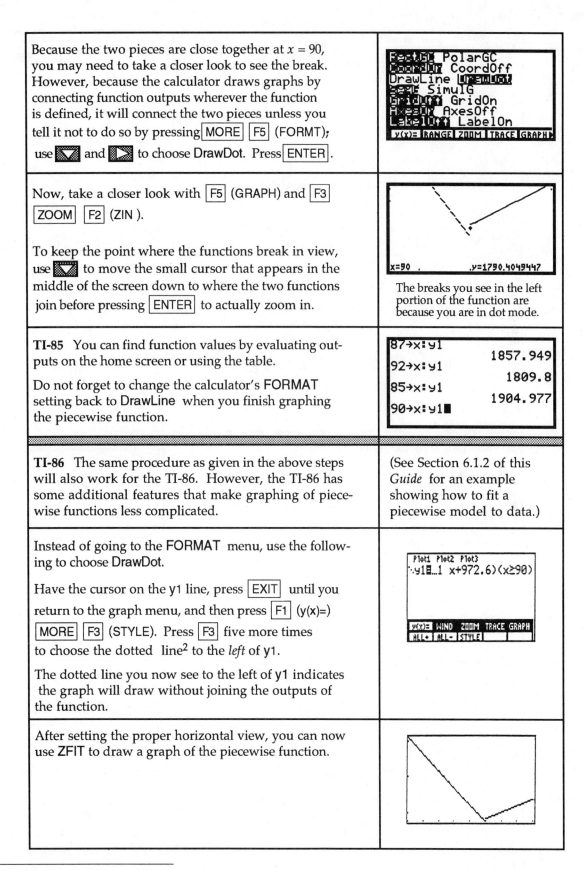

Because the two pieces are close together at $x = 90$, you may need to take a closer look to see the break. However, because the calculator draws graphs by connecting function outputs wherever the function is defined, it will connect the two pieces unless you tell it not to do so by pressing MORE F5 (FORMT); use ▼ and ▶ to choose DrawDot. Press ENTER.

Now, take a closer look with F5 (GRAPH) and F3 ZOOM F2 (ZIN).

To keep the point where the functions break in view, use ▼ to move the small cursor that appears in the middle of the screen down to where the two functions join before pressing ENTER to actually zoom in.

The breaks you see in the left portion of the function are because you are in dot mode.

TI-85 You can find function values by evaluating outputs on the home screen or using the table.

Do not forget to change the calculator's FORMAT setting back to DrawLine when you finish graphing the piecewise function.

TI-86 The same procedure as given in the above steps will also work for the TI-86. However, the TI-86 has some additional features that make graphing of piecewise functions less complicated.

(See Section 6.1.2 of this *Guide* for an example showing how to fit a piecewise model to data.)

Instead of going to the FORMAT menu, use the following to choose DrawDot.

Have the cursor on the y1 line, press EXIT until you return to the graph menu, and then press F1 (y(x)=) MORE F3 (STYLE). Press F3 five more times to choose the dotted line[2] to the *left* of y1.

The dotted line you now see to the left of y1 indicates the graph will draw without joining the outputs of the function.

After setting the proper horizontal view, you can now use ZFIT to draw a graph of the piecewise function.

---

[2]The different "graph styles" you can draw from this location are described in more detail on page 10 in your TI-86 *Graphing Calculator Guidebook.*

You can find function values by evaluating outputs on the home screen or using the table.	y1(87) 1857.949
	y1(92)
	                   1809.8
	y1(85)
	                 1904.977
	y1(90)■

📖 1.4 Linear Functions and Models

Actual real-world data is used throughout *Calculus Concepts*. It is necessary that you use your calculator to find a curve that models the data. Be very careful when you enter the data in your calculator because your model and all of your results depend on the values that you enter!

1.4.1 ENTERING DATA   There are several ways to input data in your calculator. Two of these, entering data from the home screen and entering data using the list editor, are discussed below. If you do not see the list names L1 and L2 return to the statistical setup instructions at the beginning of this *Guide*.

We will explore data entry with the following data:

Year	1992	1993	1994	1995	1996	1997
Tax	2541	3081	3615	4157	4703	5242

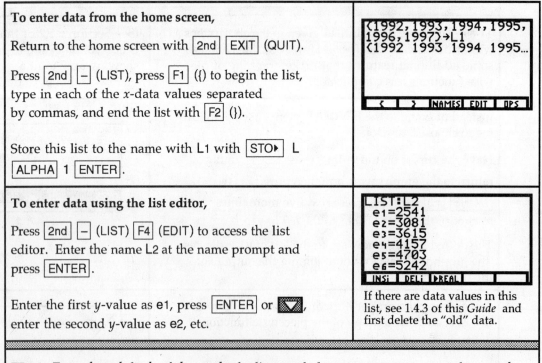

To enter data from the home screen,

Return to the home screen with [2nd] [EXIT] (QUIT).

Press [2nd] [–] (LIST), press [F1] ({) to begin the list, type in each of the *x*-data values separated by commas, and end the list with [F2] (}).

Store this list to the name with L1 with [STO▸] L [ALPHA] 1 [ENTER].

To enter data using the list editor,

Press [2nd] [–] (LIST) [F4] (EDIT) to access the list editor. Enter the name L2 at the name prompt and press [ENTER].

Enter the first *y*-value as e1, press [ENTER] or ▼, enter the second *y*-value as e2, etc.

If there are data values in this list, see 1.4.3 of this *Guide* and first delete the "old" data.

TI-86   Even though both of the methods discussed above are ways to enter data on the TI-86, the most convenient method is to enter the data in the TI-86's stat lists. Return to the home screen and press [2nd] [+] (STAT) [F2] (EDIT). Go to Section 1.4.3 of this *Guide*, delete the data currently in these lists, and enter the data again using the method described next.

Position the cursor in the first location in list L1. Enter the x-data into list L1 by typing the numbers from top to bottom in the L1 column, pressing ENTER after each entry.

After typing the L1(5) value, 1997, use 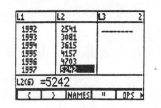 to go to the top of list L2. Enter the y-data into list L2 by typing the entries from top to bottom in the L2 column, pressing ENTER after each entry.

1.4.2 EDITING DATA IN THE TI-85 LIST EDITOR

If you incorrectly type a data value, access the data with the list editor and use the cursor keys to move to the value you wish to correct. Type the correct value and press ENTER.

- To *insert* a data value, put the cursor over the value that will be directly below the one you will insert, and press F1 (INSi). The values in the list below the insertion point move down one location and a 0 is filled in at the insertion point. Type the data value to be inserted and press ENTER. The 0 is replaced with the new value.

- To *delete* a single data value, move the cursor to the value you wish to delete, and press F2 (DELi). The values in the list below the deleted value move up one location.

1.4.2 EDITING DATA IN THE TI-86 STAT LISTS

If you incorrectly type a data value, use the cursor keys to darken the value you wish to correct and type the correct value. Press ENTER.

- To *insert* a data value, put the cursor over the value that will be directly below the one you will insert, and press 2nd DEL (INS). The values in the list below the insertion point move down one location and a 0 is filled in at the insertion point. Type the data value to be inserted over the 0 and press ENTER. The 0 is replaced with the new value.

- To *delete* a single data value, move the cursor over the value you wish to delete, and press DEL. The values in the list below the deleted value move up one location.

1.4.3 DELETING OLD DATA

Whenever you enter new data in your calculator, you should first delete any previously-entered data. There are several ways to do this, and the most convenient method is illustrated below.

TI-85 Whenever you enter new data in your calculator, you should first delete any previously-entered data using one of the following methods:

- Whenever you enter data from the home screen, previously-entered data is automatically replaced with new data. Thus, {1, 2, 3} STO▸ L2 replaces the "old" L2.

- When you enter the list editor and there is "old" data in the list, position the cursor over the e1 value and repeatedly press F2 (DELi) until all data is deleted.

TI-86   Access the data lists with [STAT] [1] (EDIT).

(You probably have different values in your lists if you are deleting "old" data.)

Use ⬆ to move the cursor over the name L1.

Press [CLEAR] [ENTER].

Use ▶ ⬆ to move the cursor over the name L2. Press [CLEAR] [ENTER].

Repeat this procedure to clear data from any of the other lists you want to use.

1.4.4 ALIGNING DATA

Let's now return to the data entered in Section 1.4.1 of this *Guide*. Suppose you want L1 to contain the number of years since a certain year (here, 1992) instead of actual years. That is, you want to *align* the x-data. In this example, you are to shift all the data values 1992 units to the left of where they currently are located.

Return to the home screen.

TI-85   Replace the L1 values with L1 – 1992 values by pressing [ALPHA] L1 [–] 1992 [STO▸] L [ALPHA] 1 [ENTER]. L1 now contains the aligned x-values.

TI-86   Position the cursor over the L1 at the top of the first column.

Replace the L1 values with L1 – 1992 values by pressing [F3] (NAMES), the menu key under L1, and then press [–] 1992 [ENTER].

Instead of an actual year, the input now represents the number of years after 1992.

1.4.5 PLOTTING DATA

Any functions you have in the y(x)= list will graph when you plot data. Therefore, you should delete them or turn them off before drawing a scatter plot.

Access the y(x)= graphing list. If any entered function is no longer needed, delete it with [F4] (DELf) or clear it with [CLEAR].

If you want the function to remain but do not want it to graph, position the cursor in that function location and press [F5] (SELCT).

A "turned off" function.

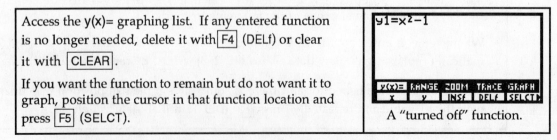

To graph data on the TI-85:

The TI-85 command Scatter L1, L2 draws a scatter plot of L1 versus L1. However, you must set an appropriate RANGE before using this command. Program STPLT found in the TI-85/TI-86 Appendix automates this task.

Press PRGM and then press F1 (NAMES). Find the name STPLT (press MORE if necessary).	Press the F-key corresponding to the location of STPLT to run the program.
Press CLEAR to remove the menu from the bottom of the screen for a better view. A scatter plot is drawn of the data with the L1 data as the input data and the L2 data as the output data.	

- Program STPLT sets the x and y-axis tick marks to 0 so they do not interfere with your view of the scatter plot.

- Because the dots the calculator uses to plot data are sometimes difficult to see when overdrawing the model of best fit, the program places a small box around each data point. (The boxes may appear a slightly different size with different screen settings.)

- Even though the TI-85 generally allows you to call lists by any names you want, *you must enter the input data in the list named L1 and the output data in the list named L2* when using program STPLT.

- Lists L1 and L2 must be of the same length or an error message results.

- It is not possible to trace a scatter plot drawn on the TI-85.

To graph data on the TI-86:

The TI-86 has a built-in command to graph data and autoscale the data window.

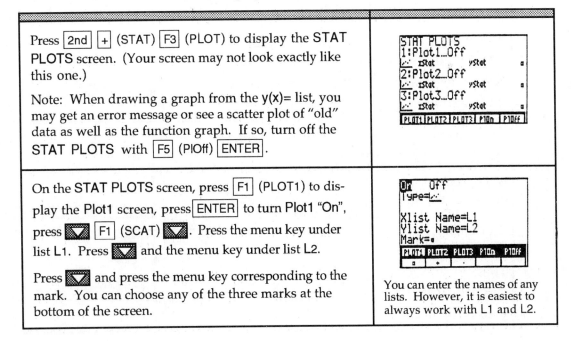

Press 2nd + (STAT) F3 (PLOT) to display the STAT PLOTS screen. (Your screen may not look exactly like this one.) Note: When drawing a graph from the y(x)= list, you may get an error message or see a scatter plot of "old" data as well as the function graph. If so, turn off the STAT PLOTS with F5 (PlOff) ENTER.	
On the STAT PLOTS screen, press F1 (PLOT1) to display the Plot1 screen, press ENTER to turn Plot1 "On", press ▽ F1 (SCAT) ▽. Press the menu key under list L1. Press ▽ and the menu key under list L2. Press ▽ and press the menu key corresponding to the mark. You can choose any of the three marks at the bottom of the screen.	You can enter the names of any lists. However, it is easiest to always work with L1 and L2.

Press EXIT until you return to the home screen. Go to the graph menu and clear the y(x)= list . To have the TI-86 set an autoscaled view of the data and draw the scatter plot, press GRAPH F3 (ZOOM) MORE F5 (ZDATA).

(ZDATA does not reset the x and y-axis tick marks. You should do this manually with RANGE or WIND if you want different spacing between the marks.)

You can trace the scatter plot with the TRACE key.

Press GRAPH and access the function list. Notice that "Plot1" at the top of the screen is now dark. This is because you have turned Plot1 "on". If you always put input data in list L1 and output data in list L2, you can turn the scatter plot off and on from the y(x)= screen rather than the stat plots screen from this point on.

To turn Plot1 off, use ▲ to move the cursor to the Plot1 position, and press ENTER . Reverse the process to turn Plot1 back on.

A scatter plot is turned *on* when its name on the y(x)= screen is darkened. Remember that you will have lower-case functions.

- TI-86 lists can be named and stored in the calculator's memory for later recall and use. If you do this and use the list by its stored name, you must use the name of the list in the stat plot setup or on the stat plot screen each time you change lists. Refer to your *TI-86 Guidebook* for details.

1.4.6 FINDING FIRST DIFFERENCES   When the input values are evenly spaced, use program DIFF to compute first differences in the output values. If the data are perfectly linear (*i.e.*, every data point falls on the graph of the line), the first differences in the output values are constant. If the first differences are "close" to constant, this is an indication that a linear model *may* be appropriate.

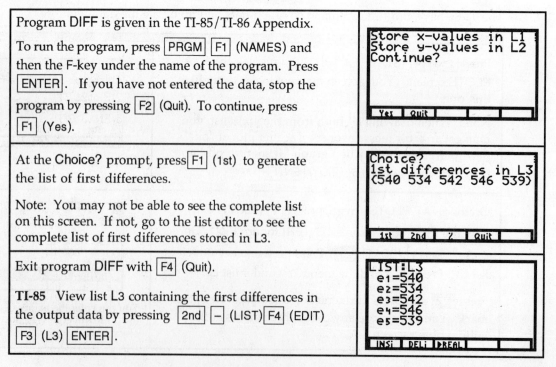

Program DIFF is given in the TI-85/TI-86 Appendix.

To run the program, press PRGM F1 (NAMES) and then the F-key under the name of the program. Press ENTER . If you have not entered the data, stop the program by pressing F2 (Quit). To continue, press F1 (Yes).

At the **Choice?** prompt, press F1 (1st) to generate the list of first differences.

Note: You may not be able to see the complete list on this screen. If not, go to the list editor to see the complete list of first differences stored in L3.

Exit program DIFF with F4 (Quit).

TI-85   View list L3 containing the first differences in the output data by pressing 2nd – (LIST) F4 (EDIT) F3 (L3) ENTER .

TI-86 If you do not want to use program DIFF, you can use your TI-86 to compute first differences of any list.

Press 2nd − (LIST) F5 (OPS) MORE MORE F4 (Delta() EXIT F3 (NAMES) and the menu key under L2. Press) ENTER to see the list of first differences in the output data.

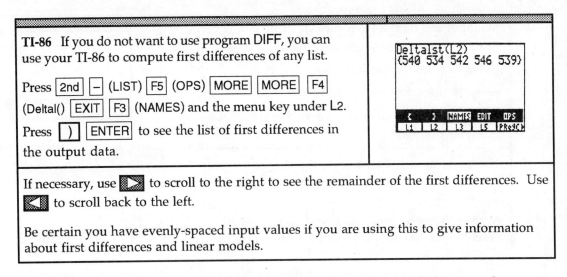

If necessary, use ▶ to scroll to the right to see the remainder of the first differences. Use ◀ to scroll back to the left.

Be certain you have evenly-spaced input values if you are using this to give information about first differences and linear models.

- **Both**   Program DIFF **should not** be used for data with input (L1) values that are *not* evenly spaced.   First differences give no information about a possible linear fit to data with inputs that are not the same distance apart.

1.4.7 FINDING A LINEAR MODEL   Use your calculator to obtain the linear model that best fits the data. Your calculator can find two different, but equivalent, forms of the linear model: $y = ax + b$ or $y = a + bx$. For convenience, we always choose the model $y = ax + b$.

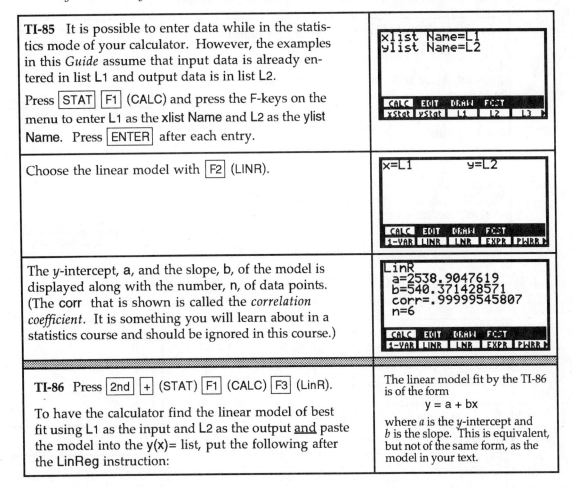

TI-85 It is possible to enter data while in the statistics mode of your calculator. However, the examples in this *Guide* assume that input data is already entered in list L1 and output data is in list L2.

Press STAT F1 (CALC) and press the F-keys on the menu to enter L1 as the xlist Name and L2 as the ylist Name. Press ENTER after each entry.

Choose the linear model with F2 (LINR).

The *y*-intercept, **a**, and the slope, **b**, of the model is displayed along with the number, **n**, of data points. (The **corr** that is shown is called the *correlation coefficient*. It is something you will learn about in a statistics course and should be ignored in this course.)

TI-86 Press 2nd + (STAT) F1 (CALC) F3 (LinR).

To have the calculator find the linear model of best fit using L1 as the input and L2 as the output and paste the model into the y(x)= list, put the following after the LinReg instruction:

The linear model fit by the TI-86 is of the form
$$y = a + bx$$
where *a* is the *y*-intercept and *b* is the slope. This is equivalent, but not of the same form, as the model in your text.

the input and output list names either by typing them from the keyboard or by pressing [2nd] [–] (LIST) [F3] (NAMES), press the menu key under L1, press [,], and press the menu key under L2. Next, type [,] y1. Press [ENTER]. (The model will be pasted into the location that you specify.)	`LinR L1,L2,y1`
	
	`CALC EDIT PLOT DRAW VARS`
	`One4a Two4a LinR LnR ExpR`

The linear model of best fit for the aligned tax data entered in Section 1.4.4 of this *Guide* is displayed on the home screen.
 	
Note: It is not necessary to first clear any previously-entered function from the location of the y(x)= list.
	`LinReg`
	` y=a+bx`
	` a=2538.90476`
	` b=540.371429`
	` corr=.999995458`
	` n=6`
	`CALC EDIT PLOT DRAW VARS`

(The **corr** that is shown is called the *correlation coefficient*. It is something you will learn about in a statistics course and should be ignored in this course.)	Go the y(x)= list to verify that the model has been pasted into the y1 location.

1.4.8 PASTING A TI-85 MODEL INTO THE FUNCTION LIST The coefficients of the
model found by the calculator should *not* be rounded. This is not a problem because the calculator will paste the entire model into the function list!

TI-85 Press [MORE] [F4] (STREG). At the **Name=** prompt, type in y1 and press [ENTER].
 	
Remember that you must use a lower-case y to refer to functions in the y(x)= graphing list.
 	
(Any function currently in y1 will be replaced with the linear model.)
	`LinR`
	` a=2538.9047619`
	` b=540.371428571`
	` corr=.99999545807`
	` n=6`
	`Name=y1`
	`CALC EDIT GRAPH FCST`
	`P2REG P3REG P4REG STREG`

Press [GRAPH] [F1] (*y(x)=*) to see the equation of the model in the y1 location. If you cannot see all of the equation, press the right arrow key to scroll the screen to the right.	`y1=2538.9047619048+5…`
	
	`y(x)= RANGE ZOOM TRACE GRAPH`
	`  x    y    INSf DELf SELCT`

1.4.9 GRAPHING A MODEL   After finding a model, you should always graph it on a
scatter plot of the data to verify that the model provides a good fit to the data.

TI-85 After you have copied the model to the y(x)= list, run program STPLT to graph the model and the scatter plot on the same screen. The model will graph first and then the scatter plot will appear. **TI-86** After viewing the model, be sure that Plot1 at the top of the y(x)= list is darkened. Then, just press [EXIT] [F5] (GRAPH) to see the model and scatter plot.	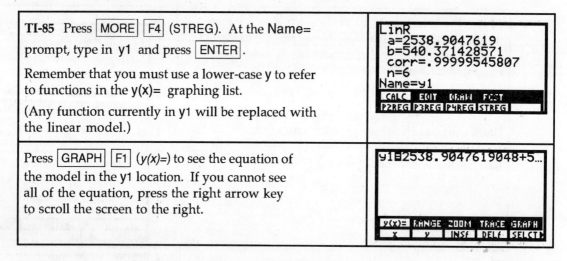

1.4.10 PREDICTIONS USING A MODEL   You could use one of the methods described in Sections 1.1.7 or 1.2.1 of this *Guide* to evaluate the linear model at the desired input value. Remember, if you have aligned your data, the input value at which you evaluate the model may not be the value given in the question you are asked.

However, your calculator has a feature that gives you a very easy way to predict either input or output values calculated from a model.

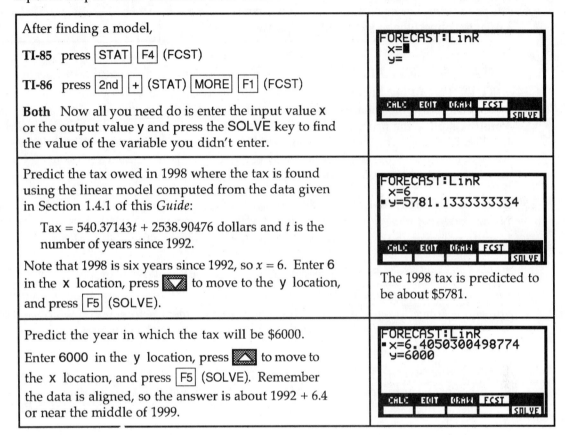

After finding a model,

TI-85 press STAT F4 (FCST)

TI-86 press 2nd + (STAT) MORE F1 (FCST)

Both Now all you need do is enter the input value x or the output value y and press the SOLVE key to find the value of the variable you didn't enter.

Predict the tax owed in 1998 where the tax is found using the linear model computed from the data given in Section 1.4.1 of this *Guide*:

Tax = $540.37143t + 2538.90476$ dollars and t is the number of years since 1992.

Note that 1998 is six years since 1992, so $x = 6$. Enter 6 in the x location, press ▼ to move to the y location, and press F5 (SOLVE).

The 1998 tax is predicted to be about $5781.

Predict the year in which the tax will be $6000.

Enter 6000 in the y location, press ▲ to move to the x location, and press F5 (SOLVE). Remember the data is aligned, so the answer is about 1992 + 6.4 or near the middle of 1999.

1.4.11 COPYING GRAPHS TO PAPER   Your instructor may ask you to copy what is on your graphics screen to paper. If so, use the following to more accurately perform this task.

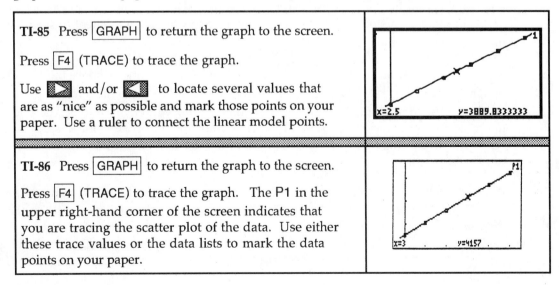

TI-85 Press GRAPH to return the graph to the screen.

Press F4 (TRACE) to trace the graph.

Use ▶ and/or ◀ to locate several values that are as "nice" as possible and mark those points on your paper. Use a ruler to connect the linear model points.

TI-86 Press GRAPH to return the graph to the screen.

Press F4 (TRACE) to trace the graph. The P1 in the upper right-hand corner of the screen indicates that you are tracing the scatter plot of the data. Use either these trace values or the data lists to mark the data points on your paper.

Both Press 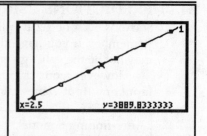 to move the trace cursor to the linear model graph. The number 1 at the top right of the screen tells you which function you are tracing (in this case, y1).

Use ▶ and/or ◀ to locate values that are as "nice" as possible and mark those points on your paper. Use a ruler to connect the model points and draw the line.

1.4.12 WHAT IS "BEST FIT"?

Even though your calculator easily computes the values a and b for the best fitting linear model $y = ax + b$, it is important to understand the method of least-squares and the conditions necessary for its application if you intend to use this model. You can explore the process of finding the line of best fit with program **LSLINE**. (Program LSLINE is given in the TI-85/TI-86 Appendix.) For your investigations of the least-squares process with this program, it is better to use data that is not perfectly linear and data for which you do *not* know the best-fitting line.

Before using program LSLINE, clear the y(x)= list and enter your data in lists L1 and L2. Next, draw a scatter plot. Reset xScl and yScl so that you can use the tick marks to help identify points on the graphics screen. Press GRAPH to view the scatter plot.

To activate program LSLINE, press PRGM F1 (NAMES) followed by the F-key under the program, and press ENTER . (Since your calculator "loads" each program in memory when you run it, longer programs such as this one take a moment before anything appears on the screen.) The program first displays the scatter plot you constructed and pauses for you to view the screen.

- While the program is calculating, there is a small vertical line in the upper-right hand corner of the graphics screen that is dashed and "moving". The program pauses several times during execution. Whenever this happens, the small vertical line is "still". You should press ENTER to resume execution.

The program next asks you to find the y-intercept and slope of *some* line you estimate will go "through" the data. (You should not expect to guess the best fit line on your first try!) After you enter a guess for the y-intercept and slope, your line is drawn and the errors are shown as vertical line segments on the graph. (You may have to wait just a moment to see the vertical line segments before again pressing ENTER .)

Next, the sum of squares of errors, SSE, is displayed for your line. Choose the TRY AGAIN? option by pressing 1 ENTER . Decide whether you want to move the y-intercept of the line or change its slope to improve the fit to the data. After you enter another guess for the y-intercept and/or slope, the process of viewing your line, the errors, and display of SSE is repeated. If the new value of SSE is smaller than the SSE for your first guess, you have improved the fit.

When it is felt that an SSE value close to the minimum value is found, you should press 2 at the TRY AGAIN? prompt. The program then overdraws the line of best fit on the graph for comparison with your last attempt and shows the errors for the line of best fit. The coefficients a and b of the best-fitting linear model $y = ax + b$ are then displayed along with the minimum SSE. Use program LSLINE to explore the method of least squares to find the line of best fit.

1.4.13 NAMING DATA LISTS (optional)   You may or may not want to use the additional features given below for data entered on your calculator. You can name data (either input, output, or both) and store it in the calculator memory for later recall.

For instance, suppose you wanted to call the following list L1 by another name :

1984   1985   1987   1990   1992

First, enter the data in the list L1. Return to the home screen. (You can view any list from the home screen by typing its name and pressing ENTER .)	L1 ⟨1984 1985 1987 1990…
Pressing ▶ allows you to scroll through the list to see the portion that is not displayed. Type L1, press STO▸ , type the letters D A T E and press ENTER to store this list with the name DATE. This list should now appear in your list menu.	L1 ⟨1984 1985 1987 1990… L1→DATE ⟨1984 1985 1987 1990… ⟨ ⟩ NAMES EDIT OPS
If you later want to access this list, press 2nd − (LIST). Under **NAMES**, find **DATE**. Press the number corresponding to the location of the list, press STO▸ , and type the location you wish to move the list to (say, L1). Press ENTER .	DATE→L1 ⟨1984 1985 1987 1990… ⟨ ⟩ NAMES EDIT OPS DATE L1 L2 L3 L4 ▸

- **TI-85** The original data remains in **DATE**. It is not deleted until you delete it using 2nd + (MEM) F2 (DELET) F4 (LIST), move the cursor with ▼ to the location of **DATE**, and press ENTER . Press 2nd EXIT (QUIT) to return to the home screen.

- **TI-86** The original data remains in **DATE**. It is not deleted until you delete it using 2nd 3 (MEM) F2 (DELET) F4 (LIST), move the cursor with ▼ to the location of **DATE**, and press ENTER . Press 2nd EXIT (QUIT) to return to the home screen.

Chapter 2 Ingredients of Change: Nonlinear Models

📖 2.1 Exponential Functions and Models

As we consider models that are not linear, it is very important that you be able to use scatter plots, numerical changes in output data, and the underlying shape of the basic functions to be able to identify which model best fits a particular set of data. Finding the model is only a means to an end -- being able to use mathematics to describe the changes that occur in real-world situations.

2.1.1 ENTERING EVENLY-SPACED INPUT VALUES (optional)

When an input list consists of many evenly-spaced values, there is a calculator command that will generate the list so that you do not have to type in the values in one by one. The syntax for this sequence command is *seq(formula, variable, first value, last value, increment)*. When entering years that differ by 1, the formula is the same as the variable and the increment is 1. Any letter can be used for the variable -- we choose to use X.

TI-85 Generate the list of years beginning with 1988, ending with 1997, and differing by 1 with:
 	
2nd − (LIST) F5 (OPS) MORE F3 (seq) x-VAR , x-VAR , 1988 , 1997 , 1) ENTER .
 	
Store the values in list L2 with STO▶ L ALPHA 2.
	`seq(x,x,1988,1997,1)`
	`(1988 1989 1990 1991…`
	`Ans→L2`
	`(1988 1989 1990 1991…`

TI-86 Access the data lists with 2nd + (STAT) F2 (EDIT). Use ◣◥ to move the cursor over the name L1. Generate the list of years beginning with 1988, ending with 1997, and differing by 1 with: F5 (OPS) MORE F3 (seq) x-VAR , x-VAR , 1988 , 1997 , 1) ENTER .	(calculator list screen showing L1 values 1988 1989 1990 1991, `L1 =seq(x,x,1988,1997,`)

2.1.2 FINDING PERCENTAGE CHANGE

When the input values are evenly spaced, use program DIFF to compute percentage change in the output values. If the data are perfectly exponential (i.e., every data point falls on an exponential model), the percentage change in the output values is constant. If the percentage change is "close" to constant, this is an indication that an exponential model *may* be appropriate.

Suppose the population of a small town between the years 1988 and 1997 is as follows:

Year	1988	1989	1990	1991	1992	1993	1994	1995	1996	1997
Population	7290	6707	6170	5677	5223	4805	4420	4067	3741	3442

Clear any old data, and enter the above data in lists L1 (year) and L2 (population). See Section 2.1.1 of this *Guide* for a convenient way to enter the years into L1.

Run program DIFF.

At the **Choice?** prompt, press F3 (%) to generate the list of percentage changes, and observe the percentage change in the output data in list L5. Exit program DIFF with F4 (Quit).

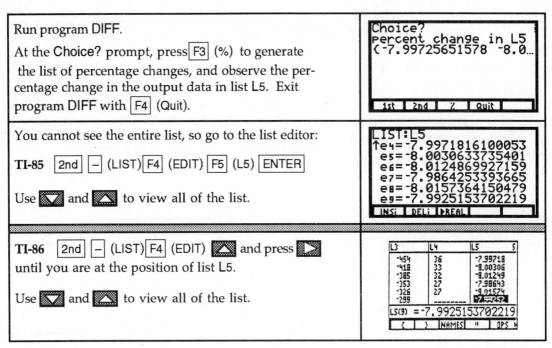

You cannot see the entire list, so go to the list editor:

TI-85   2nd   –   (LIST) F4 (EDIT) F5 (L5) ENTER

Use ▽ and △ to view all of the list.

TI-86   2nd   –   (LIST) F4 (EDIT) △ and press ▷ until you are at the position of list L5.

Use ▽ and △ to view all of the list.

- **Both**   The percentage change is very close to constant, so an exponential model may be a good fit.

2.1.3   FINDING AN EXPONENTIAL MODEL

Use your calculator to find an exponential model that fits the data. The exponential model is accessed with the statistics menu command **EXPR** on the TI-85 (**ExpR** on the TI-86) and is of the form $y = ab^x$. Using the instructions below, construct a scatter plot of the data. Notice that the data curves rather than falling in a straight line pattern.   An exponential model certainly seems appropriate!

TI-85   Use program STPLT to construct a scatter plot of the data.

TI-86   Press GRAPH F3 (ZOOM) MORE F5 (ZDATA)

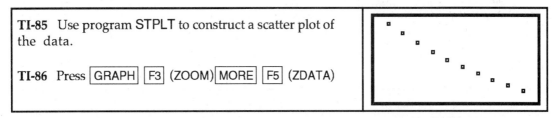

It is very important to align large numbers (like years) whenever you find an exponential model. The model found by the calculator may not even be correct if you don't!

Other alignments are possible, but we choose to align so that $x = 0$ in 1988.

Return to the home screen.

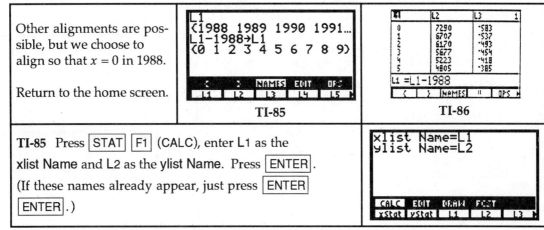

TI-85   Press STAT F1 (CALC), enter L1 as the xlist Name and L2 as the ylist Name. Press ENTER. (If these names already appear, just press ENTER ENTER.)

Choose the exponential model with $\boxed{F4}$ (EXPR). The exponential model is shown on the screen. Copy the model to the $y(x)=$ list with $\boxed{MORE}$ $\boxed{F4}$ (STREG) y1, overdraw the graph on the scatter plot with STPLT, and see that it gives a very good fit.

```
EXPR
 a=7290.25031519
 b=.919994901443
 corr=-.999999967931
 n=10

 CALC  EDIT  GRAPH  FCST
1-VAR LINR  LNR  EXPR  PWRR▸
```

TI-86 Press $\boxed{2nd}$ $\boxed{+}$ (STAT) $\boxed{F1}$ (CALC) $\boxed{F5}$ (ExpR) $\boxed{2nd}$ $\boxed{-}$ (LIST) $\boxed{F3}$ (NAMES) L1 $\boxed{,}$ L2 $\boxed{,}$ y1

Press $\boxed{ENTER}$. The model is generated and pasted into the $y(x)=$ list.

```
ExpReg
 y=a*b^x
 a=7290.25032
 b=.919994901
 corr=-.99999997
 n=10
```

TI-85 Overdraw the model on the scatter plot by running program STPLT.

TI-86 Overdraw the model on the scatter plot by using $\boxed{▲}$ to move the cursor over Plot1 in the $y(x)=$ list, press $\boxed{ENTER}$ to turn Plot1 on, and press $\boxed{EXIT}$ $\boxed{F3}$ (ZOOM) $\boxed{MORE}$ $\boxed{F5}$ (ZDATA).

2.1.4 FINDING A LOGISTIC MODEL

Use your calculator to find a logistic model of the form $y = \dfrac{L}{1 + Ae^{-Bx}}$. The logistic model that you obtain may be slightly different from a logistic model found with another calculator. Logistic models in *Calculus Concepts* were found using a TI-83. Refer to the following discussion for the comparable logistic model that best fits the data given in Example 2, Section 2.1 of the text. As with the exponential model $y = ab^x$, large input values must be aligned before fitting a logistic model to data.

Clear any old data, and enter the following in lists L1 and L2:

Aligned end of month	1	2	3	4	5	6	7	8	9
Total number of swimsuits sold	4	12	25	58	230	439	648	748	769

Construct a scatter plot of the data. A logistic model seems appropriate.

TI-85 Program LOGISTIC finds a "best-fit" logistic model rather than being a logistic model with a user-input limiting value L such that no data value is ever greater than L.

Run program[1] LOGISTIC to fit the logistic model.

Note: To use this program, the input data must be in order, from smallest to largest, in list L1. Have the output data in L2.

Run program LOGISTIC with $\boxed{PRGM}$ followed by the number of the location of the program. Press $\boxed{ENTER}$.

```
LOGISTIC




NAMES EDIT
 DIFF EULER LOGIS LSLIN MDVAL▸
```
Your program list may not look exactly like this.

---

[1]The authors express their sincere appreciation to Dr. Dan Warner and Robert Simms of the Mathematical Sciences Department at Clemson University for their invaluable help with program LOGISTIC .

The first message you see reminds you that the input data should be in list L1 and the output data in list L2. If you have not done this, press ON and choose 2 (QUIT). Enter the data and then rerun the program.	DATA IN L1,L2 ENTER CONTINUES
After pressing ENTER to continue, the program displays several messages that you can ignore. (These messages are giving information about some of the advanced calculus techniques used to fit the model. You can see SSE being reduced as the model is fit.) The program places the equation of the model in the y1 location of the y(x)= list.	MODEL Y=L/(1+A*e^(-B x)) L=786.704443572 A=1464.7016675 B=1.26109661733

TI-86 This calculator fits a logistic model of the form $y = \dfrac{a}{1+b \cdot e^{-cx}} + d$ to data. The form of this logistic model differs from those found by the TI-82, TI-83, and TI-85 because it includes d, a vertical shift from the horizontal axis. When the model is fit to data, the number of data points is displayed, and the model coefficients are in a list called **PRegC**. (The **tolMet** = 1 message concerns the tolerance of the TI-86. More information can be found on page 313 of your TI-86 *Guidebook*.)

If you would rather find the same model as the other calculators instead of using the built-in routine, program **LOGISTIC** can be transferred directly to the TI-86 from a TI-85. If so, read the TI-85 directions above for use of the program.

Fit a logistic model to the data and copy the model to the y1 location of the y(x)= list by pressing 2nd + (STAT) F1 (CALC) MORE F3 (LgstR) 2nd − (LIST) F3 (NAMES) L1 , L2 , y1 ENTER.	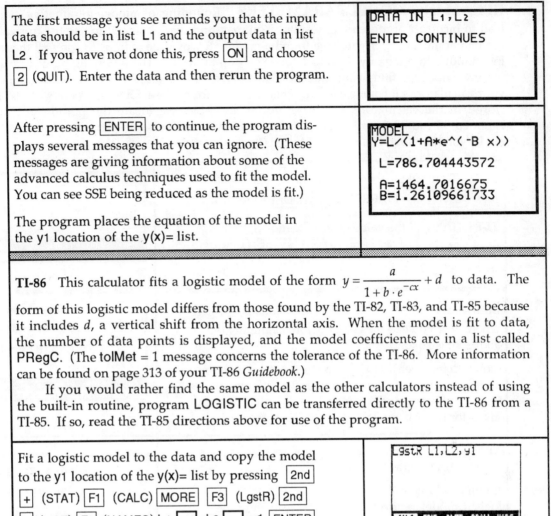
This model will take longer to generate than the other models. Notice that the TI-86 uses the variable *a* for the limiting value that your text calls *L*. The model also uses different symbols for the other parameters.	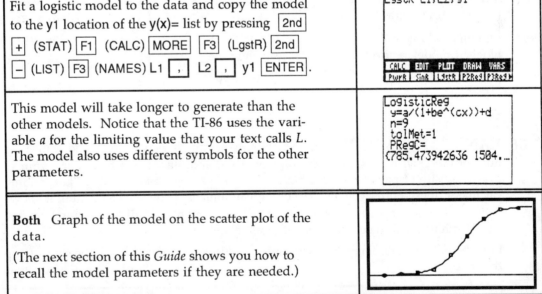
Both Graph of the model on the scatter plot of the data. (The next section of this *Guide* shows you how to recall the model parameters if they are needed.)	

- Provided the input values are evenly spaced, program DIFF might be helpful when you are trying to determine if a logistic model is appropriate for certain data. If the first differences (in list L3 after running program DIFF) *begin small*, *peak in the middle*, and *end small*, this is an indication that a logistic model may provide a good fit to the data. Such is true for this data set because the first differences are 8, 13, 33, 172, 209, 209, 100, and 21.

2.1.5 RECALLING MODEL PARAMETERS   Rounding of model parameters can often lead to incorrect or misleading results. You may find that you need to use the full values of model parameters after you have found a model. It would be tedious to copy all these digits into another location of your calculator. You don't have to! The following procedure applies for any model you find using one of the built-in regressions (*i.e.*, from the STAT CALC menu) in your calculator. Of course, once another model is found, previous parameters are no longer stored in the calculator's memory. As an example, we locate the parameter b for the exponential model found in Section 2.1.3. However, this same procedure applies to any model you have found using the STAT CALC menu.

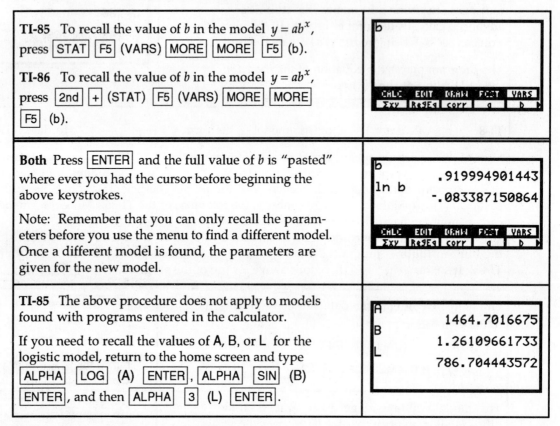

TI-85   To recall the value of b in the model $y = ab^x$, press $\boxed{\text{STAT}}$ $\boxed{\text{F5}}$ (VARS) $\boxed{\text{MORE}}$ $\boxed{\text{MORE}}$ $\boxed{\text{F5}}$ (b).

TI-86   To recall the value of b in the model $y = ab^x$, press $\boxed{\text{2nd}}$ $\boxed{+}$ (STAT) $\boxed{\text{F5}}$ (VARS) $\boxed{\text{MORE}}$ $\boxed{\text{MORE}}$ $\boxed{\text{F5}}$ (b).

Both   Press $\boxed{\text{ENTER}}$ and the full value of b is "pasted" where ever you had the cursor before beginning the above keystrokes.

Note: Remember that you can only recall the parameters before you use the menu to find a different model. Once a different model is found, the parameters are given for the new model.

TI-85   The above procedure does not apply to models found with programs entered in the calculator.

If you need to recall the values of A, B, or L for the logistic model, return to the home screen and type $\boxed{\text{ALPHA}}$ $\boxed{\text{LOG}}$ (A) $\boxed{\text{ENTER}}$, $\boxed{\text{ALPHA}}$ $\boxed{\text{SIN}}$ (B) $\boxed{\text{ENTER}}$, and then $\boxed{\text{ALPHA}}$ $\boxed{3}$ (L) $\boxed{\text{ENTER}}$.

2.1.6 RANDOM NUMBERS   Imagine all the real numbers between 0 and 1, including the 0 but not the 1, written on identical slips of paper and placed in a hat. Close your eyes and draw one slip of paper from the hat. You have just chosen a number "at random". Your calculator doesn't offer you a choice of all real numbers between 0 and 1, but it allows you to choose, *with an equal chance of obtaining each one*, any of 10^{14} different numbers between 0 and 1 with its random number generator called rand.

First, "seed" the random number generator. (This is like mixing up all the slips of paper in the hat.)

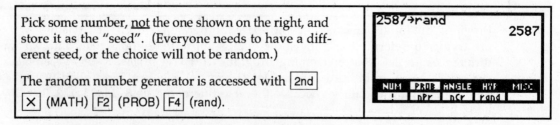

Pick some number, <u>not</u> the one shown on the right, and store it as the "seed". (Everyone needs to have a different seed, or the choice will not be random.)

The random number generator is accessed with $\boxed{\text{2nd}}$ $\boxed{\times}$ (MATH) $\boxed{\text{F2}}$ (PROB) $\boxed{\text{F4}}$ (rand).

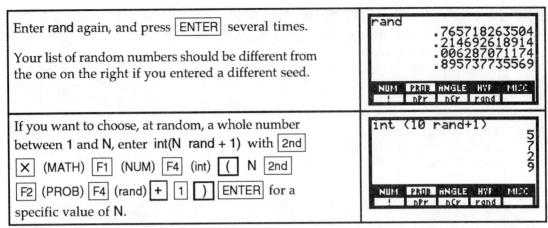

Enter **rand** again, and press ENTER several times.

Your list of random numbers should be different from the one on the right if you entered a different seed.

If you want to choose, at random, a whole number between 1 and N, enter **int(N rand + 1)** with 2nd ☒ (MATH) F1 (NUM) F4 (int) ⎡⎣(N 2nd F2 (PROB) F4 (rand) + 1)⎤ ENTER for a specific value of N.

- Repeatedly press ENTER to choose more random numbers. For instance, the screen to the right shows several values that were chosen at random with N = 10.

📖 2.2 Exponential Models in Finance

You are probably familiar with the compound interest formulas. This section introduces you to some new methods of using your calculator with familiar formulas.

2.2.1 REPLAY OF PREVIOUS ENTRIES TO FIND FORMULA OUTPUTS   You can

recall expressions previously typed by repeatedly using the calculator's last entry feature. Learn to use this time-saving feature of your calculator.

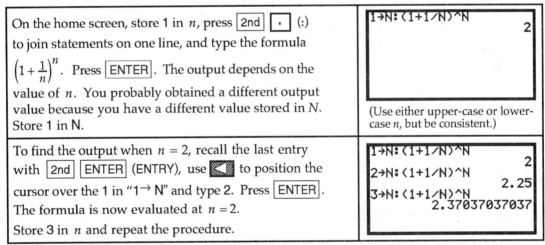

On the home screen, store 1 in n, press 2nd . (:) to join statements on one line, and type the formula $\left(1+\frac{1}{n}\right)^n$. Press ENTER. The output depends on the value of n. You probably obtained a different output value because you have a different value stored in N. Store 1 in N.

(Use either upper-case or lower-case n, but be consistent.)

To find the output when $n = 2$, recall the last entry with 2nd ENTER (ENTRY), use ◀ to position the cursor over the 1 in "1→ N" and type 2. Press ENTER. The formula is now evaluated at $n = 2$.
Store 3 in n and repeat the procedure.

- **TI-86 Note:** 2nd ENTER (ENTRY) brings back several previously-entered expressions on this calculator, so the storing and evaluating could be done in two different steps.

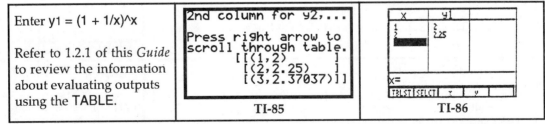

Enter y1 = (1 + 1/x)^x

Refer to 1.2.1 of this *Guide* to review the information about evaluating outputs using the TABLE.

TI-85

TI-86

Because this formula contains only one input variable, you could enter it in the y(x)= list, using x as the input variable, and find the outputs using the TABLE (as indicated above.)

2.2.2 DETERMINING FUTURE VALUE You can save a lot of keystrokes by recalling expressions previously typed by repeatedly using the calculator's last entry feature. When a formula contains more than one input variable, it's easier to recall the last entry on the home screen than to try to use the TABLE. To illustrate, consider the compound interest formula -- one that contains several input variables. Type in the formula for the amount in an account paying r% interest (compounded n times a year) on an initial deposit of $P over t years:

$$Amount = P\left(1 + \frac{r}{n}\right)^{nt} \text{ dollars}$$

Carefully watch the screen as you type the statements below. The $\boxed{\text{STO►}}$ key puts the cursor in alphabetic mode, so you must press $\boxed{\text{ALPHA}}$ before typing the colon.

The result you obtain when you evaluate the formula with $\boxed{\text{ENTER}}$ depends on the values your calculator has stored in *P*, *R*, and *T*. Store 1000 in *P*, 0.05 in *R*, 4 in *N*, and 1 in *T* and find the amount in the account to be $1050.95.	`P(1+R/N)^(N*T)` `      4.10989067286` `1000→P:.05→R:4→N:1→T:` `P(1+R/N)^(N*T)` `      1050.94533691`
Recall the last entry with $\boxed{\text{2nd}}$ $\boxed{\text{ENTER}}$ (ENTRY), use ◀ to move the cursor, and edit the statements to determine the accumulated amount if $5000 is invested at 5% interest compounded monthly for 3 years. You should find a result of $5807.36.	`       4.10989067286` `1000→P:.05→R:4→N:1→T:` `P(1+R/N)^(N*T)` `      1050.94533691` `5000→P:.05→R:12→N:3→T` `:P(1+R/N)^(N*T)` `      5807.36115667`

Because the TI-85 will only recall the last-entered expression, you should enter the single line shown above (with colons) so that it can be recalled for editing. However, the TI-86 recalls any number of previously-entered expressions, so you could store each variable separately *or* use the above procedure.

TI-86 Store the values for the various variables, enter the formula, and press $\boxed{\text{ENTER}}$. Then, store new values for the variables and repeatedly press $\boxed{\text{2nd}}$ $\boxed{\text{ENTER}}$ (ENTRY) until the formula reappears. Press $\boxed{\text{ENTER}}$ to evaluate the formula at the new values.

2.2.3 FINDING PRESENT VALUE The present value of an investment is easily found with the calculator's solver. For instance, suppose you want to solve for the present value *P* the equation $9438.40 = P\left(1 + \frac{0.075}{12}\right)^{60}$.

Refer to 1.2.2 of this *Guide* for instructions on using the TI-85's SOLVER. Enter the equation above as shown on the right.
 	
(The TI-86 F2 location is WIND, not RANGE.)
	`9438.4=P(1+.075/12)^…` `P=500` `bound=(-1ᴇ99,1ᴇ99)`
	
	`GRAPH│RANGE│ZOOM│TRACE│SOLVE`

Solve for *P* to obtain the present value $6494.49.
 	
(If you prefer, you could find the *x*-intercept of y1 = 9438.4 - x(1+.075/12)^60 to find the present value. Refer to Section 1.2.3 of this *Guide* for more detailed instructions.)
	`9438.4=P(1+.075/12)^…` `•P=6494.4858703432` `bound=(-1ᴇ99,1ᴇ99)` `•left-rt=1ᴇ-10`
	
	`GRAPH│RANGE│ZOOM│TRACE│SOLVE`

📖 2.3 Polynomial Functions and Models

You will in this section learn how to fit models to data that have the familiar shape of a parabola or a cubic. Using your calculator to find these models involves basically the same procedure as when using it to find linear and exponential models.

2.3.1 FINDING SECOND DIFFERENCES
When the input values are evenly spaced, use program DIFF to compute second differences in the output values. If the data are perfectly quadratic (*i.e.*, every data point falls on a quadratic model), the second differences in the output values are constant. If the second differences are "close" to constant, this is an indication that a quadratic model *may* be appropriate.

Clear any old data, and enter the roofing job data in lists L1 and L2:

Months after January	1	2	3	4	5	6
Number of jobs	12	14	22	37	58	84

Run program DIFF, press F2 at the Choice? prompt, and observe the second differences in list L4. The second differences are close to constant, so a quadratic model may be a good fit. Construct a scatter plot of the data. A quadratic model seems appropriate!	Choice? 2nd differences in L4 (6 7 6 5) 1st 2nd ? Quit

2.3.2 FINDING A QUADRATIC MODEL
Use your calculator to obtain a quadratic model that fits the data. The calculator's quadratic model is of the form $y = ax^2 + bx + c$ and is accessed with the command P2REG.

TI-85 Press STAT F1 (CALC) and press the F-keys on the menu to enter L1 as the xlist Name and L2 as the ylist Name. Press ENTER after each entry. (If these names already appear, just press ENTER ENTER.)	xlist Name=L1 ylist Name=L2 CALC EDIT DRAW FCST xStat yStat L1 L2 L3 ▸
Choose the quadratic model with MORE F1 (P2REG). The best fitting quadratic model is displayed. The coefficients of the model $y = ax^2 + bx + c$ are displayed in the list $\{a,b,c\}$ that can be scrolled with ▸ for viewing. Copy the model to the y(x)= list.	P2Reg n=6 PRegC= (2.92857142857 -1.12… Name=y1 CALC EDIT DRAW FCST P2REG P3REG P4REG STREG
TI-86 Press 2nd + (STAT) F1 (CALC) MORE F4 (P2Reg) 2nd − (LIST) F3 (NAMES) L1 , L2 , y1. Press ENTER. The model is generated and pasted into the y(x)= list.	QuadraticReg y=ax²+bx+c n=6 PRegC= (3.07142857143 -7.01…
Both Overdraw the graph on the scatter plot, and see that this model gives a very good fit to the data.	

2.3.3 FINDING A CUBIC MODEL

Whenever a scatter plot of the data shows a single change in concavity, a cubic or logistic model is appropriate. If a limiting value is apparent, use the logistic model. Otherwise, a cubic model should be considered. When appropriate, use your calculator to obtain the cubic model that best fits data. The calculator's cubic model is of the form $y = ax^3 + bx^2 + cx + d$ and is fit to data with the $\boxed{\text{P3REG}}$ key.

Clear any old data, and enter the average price in dollars per 1000 cubic feet of natural gas for residential use in the U.S. from 1980 through 1990 in lists L1 and L2:

Year	1980	1981	1982	1983	1984	1985	1986	1987	1988	1989	1990
Price	3.68	4.29	5.17	6.06	6.12	6.12	5.83	5.54	5.47	5.64	5.77

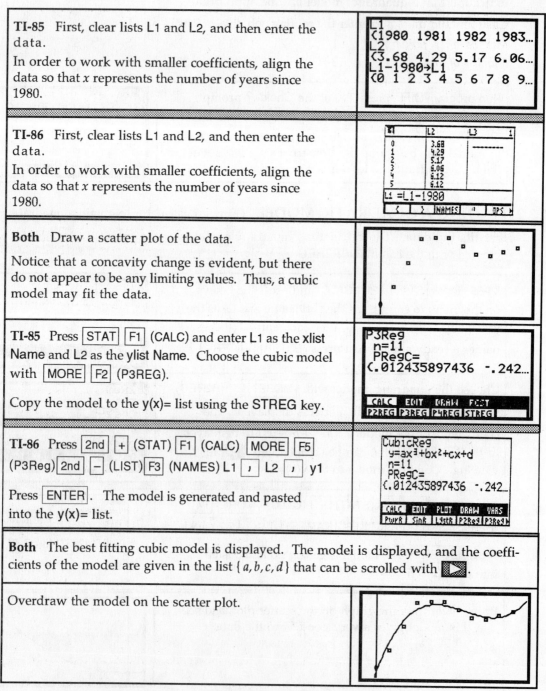

TI-85 First, clear lists L1 and L2, and then enter the data.

In order to work with smaller coefficients, align the data so that x represents the number of years since 1980.

TI-86 First, clear lists L1 and L2, and then enter the data.

In order to work with smaller coefficients, align the data so that x represents the number of years since 1980.

Both Draw a scatter plot of the data.

Notice that a concavity change is evident, but there do not appear to be any limiting values. Thus, a cubic model may fit the data.

TI-85 Press $\boxed{\text{STAT}}$ $\boxed{\text{F1}}$ (CALC) and enter L1 as the xlist Name and L2 as the ylist Name. Choose the cubic model with $\boxed{\text{MORE}}$ $\boxed{\text{F2}}$ (P3REG).

Copy the model to the y(x)= list using the STREG key.

TI-86 Press $\boxed{\text{2nd}}$ $\boxed{+}$ (STAT) $\boxed{\text{F1}}$ (CALC) $\boxed{\text{MORE}}$ $\boxed{\text{F5}}$ (P3Reg) $\boxed{\text{2nd}}$ $\boxed{-}$ (LIST) $\boxed{\text{F3}}$ (NAMES) L1 $\boxed{,}$ L2 $\boxed{,}$ y1

Press $\boxed{\text{ENTER}}$. The model is generated and pasted into the y(x)= list.

Both The best fitting cubic model is displayed. The model is displayed, and the coefficients of the model are given in the list $\{a, b, c, d\}$ that can be scrolled with $\boxed{\blacktriangleright}$.

Overdraw the model on the scatter plot.

Chapter 3   Describing Change: Rates

📖 3.1 Average Rates of Change

As you calculate average and other rates of change, remember that each numerical answer should be accompanied by units telling how the quantity is measured. You should also be able to interpret each numerical answer. It is only through their interpretations that the results of your calculations will be useful in real-world situations.

3.1.1 FINDING AVERAGE RATES OF CHANGE   Finding an average rate of change using a model is just a matter of evaluating the model at two different values of the input variable and dividing by the difference in those input values. Consider this example.

The population density of Nevada from 1950 through 1990 can be approximated by the model $P(t) = 0.1273(1.05136)^t$ people per square mile where t is the number of years since 1900. You are asked to calculate the average rates of change between from 1950 through 1980 and between 1980 and 1990.

Enter the equation in the y1 location of the y(x)= list.
 	
(Remember that you must use x as the input variable in the graphing list. You do not have to use the first function location -- any of them will do unless you intend to use program **TABLE** on the TI-85.)
 	
Return to the home screen with 2nd EXIT (QUIT).
	`y1=.1273(1.05136)^x`
	
	`Y(X)= RANGE ZOOM TRACE GRAPH`
	`  x    y    INSf  DELf  SELCT▶`

TI-85 The average rate of change of the population density between 1950 and 1980 is $\dfrac{P(80) - P(50)}{80 - 50}$.
 	
Next, evaluate the function at each of these values, store the results to different names, and then find the value of the quotient.
	`80→x:y1→A`
	`           6.99752391447`
	`50→x:y1→B`
	`           1.55740260209`
	`(A-B)/(80-50)`
	`            .181337377079`

Repeat the procedure to find the average rate of change of the population between 1980 and 1990.	`90→x:y1→A`
	`           11.546726786`
	`80→x:y1→B`
	`           6.99752391447`
	`(A-B)/(90-80)`
	`            .454920287149`

TI-86 The average rate of change of the population density between 1950 and 1980 is $\dfrac{P(80) - P(50)}{80 - 50} =$
 	
$\dfrac{y1(80) - y1(50)}{80 - 50}$. Enter this quotient, remembering to use parentheses around both the numerator and the denominator.
	`(y1(80)-y1(50))/(80-5`
	`0)`
	`            .181337377079`

To find the average rate of change between 1980 and 1990, recall the last expression with $\boxed{\text{2nd}}$ $\boxed{\text{ENTER}}$ (ENTRY) and replace the 50 by 90. Press $\boxed{\text{ENTER}}$.	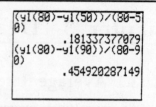

- **TI-86 Note:** If you have many average rates of change to calculate, you could put the average rate of change formula in the graphing list: $y2 = (\,y1(A) - y1(B)\,)/(A - B)$. (You, of course, need to have the model in y1.) Then, on the home screen, store the inputs of the two points in A and B: $80 \rightarrow A : 90 \rightarrow B$. All you need do then is type y1 and press enter. Store the next set of inputs into A and B and use $\boxed{\text{2nd}}$ $\boxed{\text{ENTER}}$ to recall y1 to find the average rate of change between the two new points.   Try it!

Both   Recall that rate of change units are output units per input units. We see that on average, the population density increased by about 0.18 person per square mile per year between 1950 and 1980 and by approximately 0.45 person per square mile per year between 1980 and 1990.

3.3 Tangent Lines

We first examine the principle of local linearity which says that if you are close enough, the tangent line and the curve are indistinguishable. We then use the calculator to draw tangent lines. There are two ways you can have your calculator draw a tangent line at a point on a curve. In this section, we consider one of these. The other method will be discussed in Chapter 4 of this *Guide*.

3.3.1 MAGNIFYING A PORTION OF A GRAPH   The ZOOM menu of your calculator allows you to magnify any portion of the graph of a function. Suppose we are investigating the graph of $y = {}^-x^2 + 40x + 50$ and the tangent line, $y = 20x + 150$, to the graph of this function at $x = 10$.

Enter $y = {}^-x^2 + 40x + 50$ in y1 and $y = 20x + 150$ in y2. Set the view shown to the right with RANGE on the TI-85 (or WIND on the TI-86).	RANGE xMin=0 xMax=20 xScl=1 yMin=-50 yMax=600 yScl=50 `y(x)= │RANGE│ZOOM│TRACE│GRAPH▸`
Graph the function and the line tangent to it (y1 and y2) at $x = 10$. (On the TI-86, first be certain that all plots are off.) We now want to "box in" the point of tangency and magnify that portion of the graph.	`y(x)= │RANGE│ZOOM│TRACE│GRAPH▸`
Press $\boxed{\text{F3}}$ (ZOOM) $\boxed{\text{F1}}$ (BOX) and use $\boxed{◂}$ to move the cursor to the left of and $\boxed{▾}$ to move the cursor down from the point of tangency. (You may not have the same values as those shown on the right.) Press $\boxed{\text{ENTER}}$ to fix the lower left corner of the box.	x=7.619047619¹ y=254.03225807

Use ▶ and ▲ to move the cursor to the opposite corner of your "zoom" box. Press ENTER to magnify the portion of the graph inside the box. Look at the view you now see with RANGE (or WIND). Repeat the above process if necessary.	x=12.857142857 y=453.22580645
It is easy to see that the graph of the function and the graph of the tangent line are almost the same close to the point of tangency.	MODE RANGE ZOOM TRACE GRAPH BOX ZIN ZOUT ZSTD ZPREV

- You should verify that the function and its tangent have close output values near the point of tangency by tracing the graphs near the point of tangency. Recall that you jump from one function to the other with ▲ or ▼ and that the number in the upper right-hand corner of the screen tells you on which function you are tracing.

3.3.2 DRAWING A TANGENT LINE
The GRAPH (DRAW) menu of your calculator contains the instruction to draw a tangent line to a curve at a point. To illustrate the process, we draw several tangent lines on the graph of $f(x) = x^3 + x^2 - 10x - 2$. We also investigate what the calculator does when you ask it to draw a tangent line where the line cannot be drawn.

Clear any previously-entered functions in the y(x)= list, and enter $f(x) = x^3 + x^2 - 10x - 2$ in y1.
 	
Set the view shown to the right with RANGE (or WIND). Press F5 (GRAPH).
	RANGE
	  xMin=-4.7
	  xMax=4.7
	  xScl=1
	  yMin=-30
	  yMax=30
	  yScl=10
	y(x)= RANGE ZOOM TRACE GRAPH

TI-85 Return to the home screen with 2nd EXIT (QUIT). Draw the tangent line to the curve at $x = 0$ with 2nd CUSTOM (CATALOG) – (T), use ▼ to locate TanLn(and then press ENTER. Type y1 and press , 0) ENTER.	CATALOG tan tan⁻¹ tanh tanh⁻¹ ▶TanLn(Then PAGE↓ PAGE↑ CUSTM BLANK (See box below.)
TI-86 Return to the home screen with 2nd EXIT (QUIT). Draw the tangent line to the curve at $x = 0$ with 2nd CUSTOM (CATALG-VARS) F1 (CATLG) – (T), use ▼ to locate TanLn(and then press ENTER. Type y1 and press , 0) ENTER.	TanLn(y1,0) (See box above.)
Both Notice that the tangent line cuts through the curve at $x = 0$. It appears that (0, ⁻2) is an inflection point.	x=0 y=-2

| Return to the home screen, and recall the last entry with $\boxed{\text{2nd}}$ $\boxed{\text{ENTER}}$ (ENTRY). Edit the statement so that you can draw the tangent line at $x = {}^-3$. | ```
TanLn(y1,0)
TanLn(y1,-3)
``` |
| Once again recall the last entry on the home screen, and then draw the tangent line at $x = 1.5$.

The tangent line is almost, but not quite, horizontal at $x = 1.5$. | |

Let us now look at some special cases:

1. What happens if the tangent line is vertical? We consider the function $f(x) = (x + 1)^{1/3}$ which has a vertical tangent at $x = {}^-1$.

2. How does the calculator respond when the tangent line cannot be drawn at a point? We illustrate what happens with $g(x) = |x| - 1$, a function that has a sharp point at $(0, {}^-1)$.

3. Does the calculator draw the tangent line at the joining point(s) of a piecewise continuous function? We consider two situations:

   a. $h(x)$, a piecewise continuous function that is continuous at all points and

   b. $m(x)$, a piecewise continuous function that is not continuous at $x = 1$.

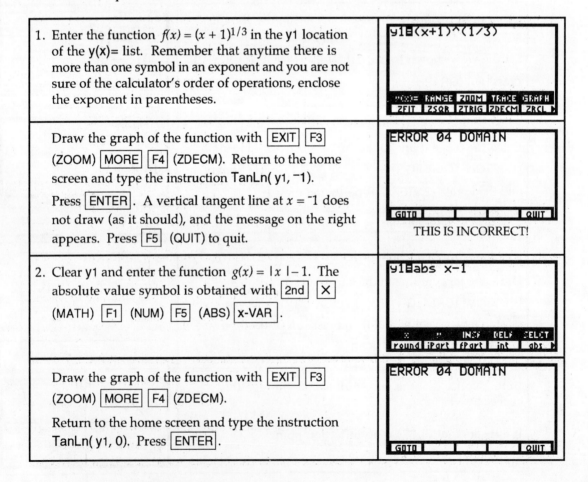

1. Enter the function $f(x) = (x + 1)^{1/3}$ in the y1 location of the y(x)= list. Remember that anytime there is more than one symbol in an exponent and you are not sure of the calculator's order of operations, enclose the exponent in parentheses.	
Draw the graph of the function with $\boxed{\text{EXIT}}$ $\boxed{\text{F3}}$ (ZOOM) $\boxed{\text{MORE}}$ $\boxed{\text{F4}}$ (ZDECM). Return to the home screen and type the instruction $\text{TanLn}(\text{y1}, {}^-1)$. Press $\boxed{\text{ENTER}}$. A vertical tangent line at $x = {}^-1$ does not draw (as it should), and the message on the right appears. Press $\boxed{\text{F5}}$ (QUIT) to quit.	
2. Clear y1 and enter the function $g(x) = |x| - 1$. The absolute value symbol is obtained with $\boxed{\text{2nd}}$ $\boxed{\times}$ (MATH) $\boxed{\text{F1}}$ (NUM) $\boxed{\text{F5}}$ (ABS) $\boxed{\text{x-VAR}}$.	
Draw the graph of the function with $\boxed{\text{EXIT}}$ $\boxed{\text{F3}}$ (ZOOM) $\boxed{\text{MORE}}$ $\boxed{\text{F4}}$ (ZDECM). Return to the home screen and type the instruction $\text{TanLn}(\text{y1}, 0)$. Press $\boxed{\text{ENTER}}$.	
```

This error message is correct! There is a sharp point at $(0, {}^-1)$, and the limiting positions of secant lines from the left and the right of that point are different. A tangent line cannot be drawn at $(0, {}^-1)$ because the instantaneous rate of change at that point does not exist.

3a. Clear y1 and enter, as indicated, the function $$h(x) = \begin{cases} x^2 & \text{when } x \le 1 \\ x & \text{when } x > 1 \end{cases}$$ [Recall that the inequality symbols are accessed with $\boxed{\text{2nd}}$ $\boxed{2}$ (TEST)].	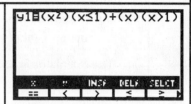 $h(x)$ is continuous for all values of x.
Draw the graph of the function with $\boxed{\text{EXIT}}$ $\boxed{\text{F3}}$ (ZOOM) $\boxed{\text{MORE}}$ $\boxed{\text{F4}}$ (ZDECM). Return to the home screen and enter TanLn(y1, 1). The calculator is correct -- the tangent line can not be drawn because secant lines drawn with points on the right and left of $x = 1$ do not approach the same slope.	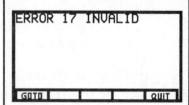
3b. Edit y1 to enter, as indicated, the function $$m(x) = \begin{cases} x^2 & \text{when } x \le 1 \\ x + 1 & \text{when } x > 1 \end{cases}$$	
TI-85 Press $\boxed{\text{EXIT}}$ $\boxed{\text{MORE}}$ $\boxed{\text{F3}}$ (FORMT) and choose DrawDot.	
TI-86 Have the cursor on the first line of the function, press $\boxed{\text{F1}}$ (y(x)=) $\boxed{\text{MORE}}$ $\boxed{\text{F3}}$ (STYLE), and press $\boxed{\text{F3}}$ five more times to choose the dotted line to the *left* of y1.	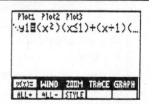
Both Draw the graph of the function with $\boxed{\text{EXIT}}$ $\boxed{\text{F3}}$ (ZOOM) $\boxed{\text{MORE}}$ $\boxed{\text{F4}}$ (ZDECM). Since $m(x)$ is not continuous at $x = 1$, the instantaneous rate of change does not exist at that point. The tangent line cannot be drawn at $(1, 1)$.	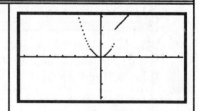

TI-85 Press GRAPH MORE F3 (FORMT) and return your calculator to Connected mode.	**TI-86** Press F1 (y(x)=) MORE F3 (STYLE) to return the slanted line to normal graphing or clear the function.

Both Return to the home screen and type the instruction TanLn(y1, 1). Press ENTER . The calculator is correct; a tangent line cannot be drawn when $x = 1$.	ERROR 17 INVALID GOTO ⬜ QUIT

Caution: Be certain that the instantaneous rate of change exists at a point before using your calculator to draw a tangent line at that point. Because of the way your calculator computes instantaneous rates of change, it may draw a tangent line at a point on a curve where the tangent line does not exist. If you receive an error message, be certain you understand why that message is the result of your action.

📖 3.5 Percentage Change and Percentage Rates of Change

The calculations in this section involve no new calculator techniques. When calculating percentage change or percentage rates of change, you have the option of using a program or the home screen.

3.5.1 CALCULATING PERCENTAGE CHANGE

Recall that program DIFF stores percentage changes (also called percentage differences) in output data in list L5. Consider the following data giving quarterly earnings for a business:

Quarter ending	Mar 1994	June 1994	Sept 1994	Dec 1994	Mar 1995	June 1995
Earnings (millions)	27.3	28.9	24.6	32.1	29.4	27.7

First, we enter the data in the calculator's lists L1 and L2.

TI-85 Align the input data so that x is the number of quarters since March 1994. Input x in L1 and earnings (in millions) in L2.	{0,1,2,3,4,5}→L1 {0 1 2 3 4 5} {27.3,28.9,24.6,32.1, 29.4,27.7}→L2 {27.3 28.9 24.6 32.1… ⬛ ⬛ NAMES EDIT OPS L1 L2 L3 L4 L5
Run program DIFF and view the percentage change in list L5. Notice that the percentage change from the end of September 1994 through December 1994 is about 30.5%. Also, from the end of March 1995 through June 1995, the percentage change is approximately ⁻5.8%.	LIST:L5 e1=5.8608058608059 e2=⁻14.878892733564 e3=30.487804878049 e4=⁻8.411214953271 e5=⁻5.7823129251701 INSi DELi ▶REAL
TI-86 Align the input data so that x is the number of quarters since March 1994. Input x in L1 and earnings (in millions) in L2.	L1 L2 L3 2 0 27.3 -------- 1 28.9 2 24.6 3 32.1 4 29.4 5 27.7 L2(6) =27.7 { } NAMES " OPS

Run program DIFF and view the percentage change in list L5.
 	
Notice that the percentage change from the end of September 1994 through December 1994 is about 30.5%. Also, from the end of March 1995 through June 1995, the percentage change is approximately ⁻5.8%.
	```L3 IL4 L5 5```
	 ```1.6 -5.9 5.860806```
	 ```-4.3 11.9 -14.9799```
	 ```7.5 -10.2 30.4879```
	 ```-2.7 1 -9.41121```
	 ```-1.7 ------- -5.78231```
	 ```L5 =(5.86080586081, -1...```
	 ```( ) NAMES| " |PS H```

Both You may find it easier to calculate these using the percentage change formula than have the program do it for you.	```(32.1-24.6)/24.6:Ans*```
	 ```100```
	 ``` 30.487804878```
	 ```(27.7-29.4)/29.4:Ans*```
	 ```100```
	 ``` -5.78231292517```

3.5.2 CALCULATING PERCENTAGE RATE OF CHANGE   Consider again the quarterly earnings for a business. Suppose you are told or otherwise find that the rate of change at the end of the June 1994 is 1.8 million dollars per quarter. Evaluate the percentage rate of change at the end of June 1994.

Divide the rate of change at the end of June 1994 by the earnings, in millions, at the end of June 1994 and multiply by 100 to obtain the percentage rate of change at that point.
 	
The percentage rate of change in earnings at the end of June 1994 was approximately 6.2% per quarter.
	```1.8/28.9```
	 ``` .062283737024```
	 ```Ans*100```
	 ``` 6.22837370242```

Chapter 4   Determining Change: Derivatives

📖 4.1 Numerically Finding Slopes

Using your calculator to find slopes of tangent lines does not involve a new procedure. However, the techniques in this section allow you to repeatedly apply a method of finding slopes that gives quick and accurate results.

4.1.1 NUMERICALLY INVESTIGATING SLOPES ON THE HOME SCREEN

Finding slopes of secant lines joining the point at which the tangent line is drawn to increasingly close points on a function to the left and right of the point of tangency is easily done using your calculator. Suppose we want to find the slope of the tangent line at $t = 8$ to the graph of the function giving the number of polio cases in 1949: $y = \dfrac{42183.911}{1 + 21484.253e^{-1.248911t}}$ where $t = 1$ on January 31, 1949, $t = 2$ on February 28, 1949, and so forth.

Enter the equation in the y1 location of the y(x)= list. (Carefully check the entry of your equation, especially the location of the parentheses. Parentheses are needed around the denominator and around the exponent.)
 	
 We now evaluate the slopes joining nearby points to the *left* of $x = 8$.
	`y1□...53e^(-1.248911x))`
	
	 `y(x)=` `RANGE` `ZOOM` `TRACE` `GRAPH`
	 `x` `y` `INSf` `DELf` `SELCT▶`

TI-85 Type in the expressions shown to the right to compute the slope of the secant line joining $x = 7.9$ and $x = 8$. Carefully watch the parentheses!
 	
 Record each slope on your paper as it is computed. You are trying to find what these slopes are approaching.
	`7.9→x:y1→A:8→x:y1→B:(`
	 `A-B)/(7.9-8)`
	 `            13159.6827248`

Press ⟨2nd⟩ ⟨ENTER⟩ (ENTRY) to recall the last entry, and then use the cursor keys to move the cursor over the 9 in the "7.9". Press ⟨2nd⟩ ⟨DEL⟩ (INS) and press ⟨9⟩ to insert another 9 in <u>both</u> positions where 7.9 appears. Press ⟨ENTER⟩ to find the slope of the secant line joining $x = 7.99$ and $x = 8$.)	`7.9→x:y1→A:8→x:y1→B:(`
	 `A-B)/(7.9-8)`
	 `            13159.6827248`
	 `7.99→x:y1→A:8→x:y1→B:`
	 `(A-B)/(7.99-8)`
	 `            13170.6176627`

Continue in this manner, recording each result, until you can determine to which value the slopes from the left seem to be getting closer and closer.	`            13170.6176627`
	 `7.999→x:y1→A:8→x:y1→B`
	 `:(A-B)/(7.999-8)`
	 `            13170.187131`
	 `7.9999→x:y1→A:8→x:y1→`
	 `B:(A-B)/(7.9999-8)`
	 `            13170.12882`

We now evaluate the slopes joining nearby close points to the *right* of $x = 8$.
 	
 Clear the screen with ⟨CLEAR⟩, recall the last expression with ⟨2nd⟩ ⟨ENTER⟩ (ENTRY), and edit it with ⟨DEL⟩ so that the nearby point is $x = 8.1$. Press ⟨ENTER⟩.
	`8.1→x:y1→A:8→x:y1→B:(`
	 `A-B)/(8.1-8)`
	 `            13146.3842089`

Continue in this manner as before, recording each result on paper, until you can determine the value the slopes from the right seem to be approaching. When the slopes from the left and the slopes from the right approach the same value, that value is the slope of the tangent line at $x = 8$.	```8.1→x:y1→A:8→x:y1→B:(A-B)/(8.1-8) 13146.3842089```
TI-86 Type in the expression shown to the right to compute the slope of the secant line joining $x = 7.9$ and $x = 8$. You must use parentheses around both the numerator and the denominator of the slope formula. Record each slope on your paper as it is computed. You are trying to find what these slopes are approaching.	```(y1(7.9)-y1(8))/(7.9-8) 13159.6827248```
Press 2nd ENTER (ENTRY) to recall the last entry, and then use the cursor keys to move the cursor over the 9 in the "7.9". Press 2nd DEL (INS) and press 9 to insert another 9 in <u>both</u> positions where 7.9 appears. Press ENTER to find the slope of the secant line joining $x = 7.99$ and $x = 8$.	```(y1(7.9)-y1(8))/(7.9-8) 13159.6827248(y1(7.99)-y1(8))/(7.99-8) 13170.6176627```
Continue in this manner, recording each result, until you can determine to which value the slopes from the left seem to be getting closer and closer.	``` 13170.6176627(y1(7.999)-y1(8))/(7.999-8) 13170.187131(y1(7.9999)-y1(8))/(7.9999-8) 13170.12882```
We now evaluate the slopes joining nearby close points to the *right* of $x = 8$. Clear the screen with CLEAR , recall the last expression with 2nd ENTER (ENTRY), and edit it with DEL so that the nearby point is $x = 8.1$. Press ENTER .	```(y1(8.1)-y1(8))/(8.1-8) 13146.3842089```
Continue in this manner as before, recording each result on paper, until you can determine the value the slopes from the right seem to be approaching. When the slopes from the left and the slopes from the right approach the same value, that value is the slope of the tangent line at $x = 8$.	``` 13169.2843875(y1(8.001)-y1(8))/(8.001-8) 13170.053799(y1(8.0001)-y1(8))/(8.0001-8) 13170.11549```

Both The slopes from the left and from the right appear to be getting closer and closer to 13,170. (The number of polio cases makes sense only as a whole number.)

4.1.2 NUMERICALLY INVESTIGATING SLOPES USING THE TI-86 TABLE

The process shown in Section 4.1.1 can be done in fewer steps when you use the **TABLE**. Recall that we are evaluating the slope formula

$$\frac{f(x + h) - f(x)}{(x + h) - x} = \frac{f(8 + h) - f(8)}{h}$$

for various values of h where h is the distance from 8 to the input of the close point. This process is illustrated using the logistic function given in Section 4.1.1 of this *Guide*.

TI-86 Remember that when in the graphing list, you must use x as the input variable. Since h is what is varying in the slope formula, replace h by **x** and enter the slope formula in **y2**. Turn **y1** off since we are looking only at the output from **y2**.	Plot1 Plot2 Plot3 \y1=42183.911/(1+214… \y2▪(y1(8+x)−y1(8))/x x²y³ WIND ZOOM TRACE GRAPH x | y | INSF | DELF | SELCT▸
Press [TABLE] [F2] (TBLST) and choose the **ASK** setting. (Since we are using the **ASK** feature, the settings for **TblStart** and **ΔTbl** do not matter.)	TABLE SETUP TblStart=0 ΔTbl=1 Indpnt: Auto **ASK** TABLE
Access the table with [F1] (TABLE), and either delete or type over any previous entries in the **x** column. Let **x** (really h) take on values that move the nearby point on the left closer and closer to 8.	x | y2 ‑.1 | 13153.68 ‑.01 | 13170.62 ‑.001 | 13170.19 ‑1E‑4 | 13170.13 ‑1E‑5 | 13170.12 ‑1E‑6 | 13170.12 y2=13170.123 TBLST SELCT | x | y

- Notice that after a certain point, the calculator switches your input values to scientific notation and displays rounded output values so that the numbers can fit on the screen in the space allotted for outputs of the table. You should position the cursor over each output value and record on paper as many decimal places as necessary in order to determine the limit from the left to the desired degree of accuracy.

Repeat the process, letting **x** (really h) take on values that move the nearby point on the right closer and closer to 8. View the entire decimal value for each output and determine the limit from the right.	x | y2 .1 | 13146.38 .01 | 13169.28 .001 | 13170.05 1E‑4 | 13170.12 1E‑5 | 13170.12 1E‑6 | 13170.12 y2=13170.121 TBLST SELCT | x | y

4.1.3 VISUALIZING THE LIMITING PROCESS (optional)   Program SECTAN can be used to view secant lines between a point $(a, f(a))$ and some close points on a curve $y = f(x)$ and the tangent line at the point $(a, f(a))$. Using this program either before or after numerically finding the limit of the slopes can help you understand the numerical process.

We use the function giving the number of polio cases in 1949: $y = \dfrac{42183.911}{1 + 21484.253e^{-1.248911t}}$

where $t = 1$ on January 31, 1949, $t = 2$ on February 28, 1949, and so forth.

(Program SECTAN is given in the TI-85/86 Appendix and should be in your calculator before you work through the following illustration.)

Before using program SECTAN, the function must be in the y1 location of the y(x)= list, and you *must* draw a graph of the function. Enter the function, using *x* as the input variable, in y1. If the function is already entered, be certain it is turned on.	

Since *t* = 1 represents January 31, 1949, 0 represents the beginning of 1949. The function gives the number of polio cases for the entire year, so we view the graph through December 31, 1949 (*t* = 12).

When you need to draw the graph of a function, you usually are given the input values in the statement of the problem in your text. *Always carefully read the problem before starting the solution process.*

Press [RANGE] (WIND on the TI-86), set xMin to 0, and set xMax to 12. Press [ZOOM] [MORE] [F1] (ZFIT). Reset yMin to ‾6000 to allow room to see the trace cursor. Press [F5] (GRAPH).	RANGE xMin=0 xMax=12 xScl=1 yMin=‾6000 yMax=41904.8860405 yScl=10000 y(x)= RANGE ZOOM TRACE GRAPH▸
The graph of the function is displayed. If the point of tangency is not such that you can easily see points to the left and right of it, adjust the window settings. Return to the home screen.	x=8 y=21262.928133
Press [PRGM] (NAMES) and the F-key corresponding to the location of program SECTAN. (Your program list may not look like the one on the right.) Press [ENTER].	SECTAN NAMES EXIT NWINT SECTA SINRE STPLT TABLE▸
If you start this program and have forgotten to enter the function in y1 or to draw a graph of the function, press [F2] (No). Otherwise, press [F1] (Yes). At the prompt, enter the input value of the tangent point. (For this example, *x* = 8.)	Have f(x) in y1 and draw graph of f Continue? x-value of point of tangency? 8
The next message that appears tells you to press enter to see secant lines drawn between the point of tangency and close points to the *left*. Press [ENTER]. (Five secant lines will draw ; they may take a little time to draw.)	
When you finish looking at the graph of the secant lines, press [ENTER] to continue. The next message that appears tells you to press enter to see secant lines drawn between the point of tangency and close points to the *right*. Press [ENTER].	

Press ENTER to continue the program.

You are next instructed to press ENTER to see a graph of the tangent line. (Note that the tangent line cuts through the graph because an inflection point occurs at $x \approx 8$.)

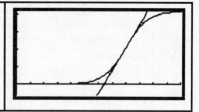

- *Caution:* In order to properly view the secant lines and the tangent line, it is essential that you first draw a graph of the function clearly showing the function, the point of tangency, and enough space so that the close points on either side can be seen.

📖 4.3 Slope Formulas

Your calculator can draw slope formulas. However, to do so, you must first enter a formula for the function whose slope formula you want the calculator to draw. Because you will probably be asked to draw slope formulas for functions whose equations you are not given, you must not rely on your calculator to do this for you. You should instead use technology to check your hand-drawn graphs and to examine the relationships between a function graph and its slope graph. It is very important in both this chapter and several later chapters that you know these relationships.

4.3.1 UNDERSTANDING YOUR CALCULATOR'S SLOPE FUNCTION   Both the TI-85 and the TI-86 use the slope of a secant line to approximate the slope of the tangent line at a point on the graph of a function. However, instead of using a secant line through the point of tangency and a close point, these calculators use the slope of a secant line through two close points that are equally spaced from the point of tangency.

Figure 7 illustrates the secant line joining the points *(a-k, f(a-k))* and *(a+k, f(a+k))*. Notice that the slopes appear to be close to the same value even though the secant line is not the same line as the tangent line.

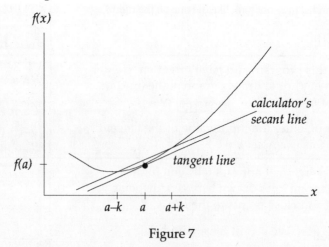

Figure 7

As *k* gets closer and closer to 0, the two points move closer and closer to *a*. Provided the slope of the tangent line exists, the limiting position of the secant line will be the tangent line.

The calculator's notation for the slope of the secant line shown in Figure 7 is

    nDer(function, symbol for input variable, *a*)

The *k* that determines the location of the points *(a-k, f(a-k))* and *(a+k, f(a+k))* can be set to different values, but specifying this value is optional. If it is not given, the calculator automatically uses $k = 0.001$ (or whatever the current setting happens to be).

Your calculator also computes for many functions[1] the slope of the tangent line. This tangent line slope, the calculator's first derivative der1, uses the syntax

der1(function, symbol for input variable, a)

We now investigate these ideas and compare nDer and der1. We begin our discussion with the smooth, continuous function $f(x) = x^3 - 4x^2 + 3.27x - 8.65$.

Enter $f(x) = x^3 - 4x^2 + 3.27x - 8.65$ in one of the locations of the y(x)= list, say y1.
 	
Return to the home screen with [2nd] [EXIT] (QUIT).
 	
Suppose you want to find the slope of the secant line between the points $(^-1, f(^-1))$ and $(1, f(1))$. That is, you are finding the slope of the secant line between $(a-k, f(a-k))$ and $(a+k, f(a+k))$ for $a = 0$ and $k = 1$.
	```y1⊟…-4 x²+3.27 x-8.65```
	
	```y(x)= RANGE ZOOM TRACE GRAPH```
	``` x y INSf DELf SELCT▸```

Your calculator uses the symbol δ for k. (The "tol" sets the calculator's tolerance -- a quantity that can affect the speed and accuracy of some computations. For our purposes, have tol $= 10^{\wedge}{-}6 = 1\text{E}^-6$.) Choose $k = 1$.
 	
TI-85 Set the value of k with [2nd] [CLEAR] (TOLER)
 	
TI-86 Set the value of k with [2nd] [3] (MEM) [F4] (TOL)
	```TOLERANCE```
	```tol=1E-6```
	```δ=1```
	
	$\delta = k$

Both Return to the home screen. Type the expressions on the right. Access nDer(with [2nd] [÷] (CALC) [F2] and with der1 with [2nd] [÷] (CALC) [F3].
 	
Press [ENTER] and see that the slope of the secant line between $(^-1, f(^-1))$ and $(1, f(1))$ is 4.27 and the slope of the tangent line at $(0, f(0))$ is 3.27.
	```nDer(y1,x,0)```
	``` 4.27```
	```der1(y1,x,0)```
	``` 3.27```
	
	```evalF nDer der1 der2 fnInt▸```

Go to the **TOLERANCE** screen and change the value of k from 1 to 0.1. Find the slope of the secant line and the slope of the tangent line.
 	
Repeat the process for $k = 0.001$ and $k = 0.0001$. Did you get the values 3.28, 3.2701, 3.270001, and 3.27000001 for the secant line slopes? Note that the value of der1 has remained constant at 3.27.
	``` 3.27```
	```nDer(y1,x,0)```
	``` 3.2701```
	```nDer(y1,x,0)```
	``` 3.270001```
	```nDer(y1,x,0)```
	``` 3.27000001```

In the table on the next page, the first row lists some values of a, the input of the point of tangency, and the second row gives the slope of the tangent line at those values. (You will later learn how to find these exact values of the slope of the tangent line to $f(x)$ at various input values.)

Use your calculator to verify the values in the third through sixth rows that give the values of der1 and nder, the slope of the secant line between the points $(a-k, f(a-k))$ and $(a+k, f(a+k))$ for the indicated values of k. Find each secant line slope by changing the value of k on the tolerance screen and then calculating the value of nDer(y1, X, a).

---

[1]The first derivative, der1, and the second derivative, der2, are valid for single-argument functions. For more information, consult your TI-85 or TI-86 owner's *Guidebook*.

a = input of point of tangency	-1	2.3	12.82	62.7
slope of tangent line	14.27	0.74	393.7672	11295.54
value of der1(y1,x,a)	14.27	0.74	393.7672	11295.54
slope of secant line, $k = 0.1$	14.28	0.75	393.7772	11295.55
slope of secant line, $k = 0.01$	14.2701	0.7401	393.7673	11295.5401
slope of secant line, $k = 0.001$	14.270001	0.740001	393.767201	11295.54
slope of secant line, $k = 0.0001$	14.27000001	0.74000001	393.7672	11295.54

You can see that the value of der1(y1,x,a) equals the slope of the tangent line and that the slope of the secant line is very close to the slope of the tangent line for small values of k. The slope of this secant line does a good job of approximating the slope of the tangent line when k is very small. For consistency, we choose $k = 0.001$. Set k to this value and do not change it.

Will the slope of this secant line always do a good job of approximating the slope of the tangent line when k is very small? Yes, it does, as long as the instantaneous rate of change exists at the input value (a) at which you evaluate nDer. When the instantaneous rate of change does not exist at a point, neither nDer nor der1 should be used to approximate or find something that does not have a value! Consider the following.

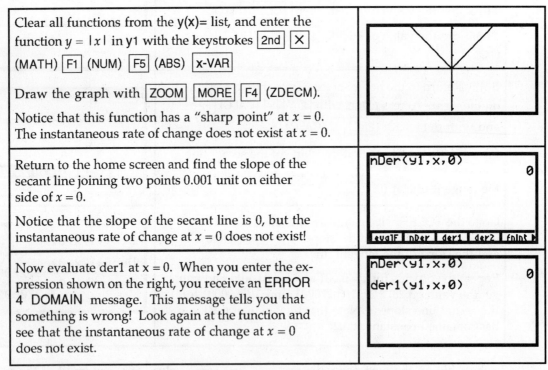

Clear all functions from the y(x)= list, and enter the function $y = \|x\|$ in y1 with the keystrokes 2nd X (MATH) F1 (NUM) F5 (ABS) x-VAR Draw the graph with ZOOM MORE F4 (ZDECM). Notice that this function has a "sharp point" at $x = 0$. The instantaneous rate of change does not exist at $x = 0$.	
Return to the home screen and find the slope of the secant line joining two points 0.001 unit on either side of $x = 0$. Notice that the slope of the secant line is 0, but the instantaneous rate of change at $x = 0$ does not exist!	nDer(y1,x,0) 0 evalF nDer der1 der2 fnInt
Now evaluate der1 at $x = 0$. When you enter the expression shown on the right, you receive an ERROR 4 DOMAIN message. This message tells you that something is wrong! Look again at the function and see that the instantaneous rate of change at $x = 0$ does not exist.	nDer(y1,x,0) 0 der1(y1,x,0)

- Be certain the instantaneous rate of change exists at a point before using nDer. Two places where nDer usually does *not* give correct results *for the instantaneous rate of change* are at sharp points and the joining point(s) of piecewise continuous functions.

- Provided the instantaneous rate of change exists at a point, we can use the secant line slope nDer to provide a good *approximation* to the slope of the tangent line at that point or we can use der1 to give the *value* of the slope of the tangent line at that point. Since the slope of the tangent line is the slope of the curve which is the derivative of the function, we choose to use der1 rather than nDer, and we call der1 the calculator's *numerical derivative*.

4.3.2 DERIVATIVE NOTATION AND CALCULATOR NOTATION

You can often see a pattern in a table of values for the slopes of a function at indicated values of the input variable and discover a formula for the slope (derivative). The process of calculating the slopes uses the calculator's numerical derivative, der1(f(X), X, X). The correspondence between our notation $\dfrac{d\,f(x)}{dx}$ and the calculator's notation der1($f(x)$, x, x) is shown below:

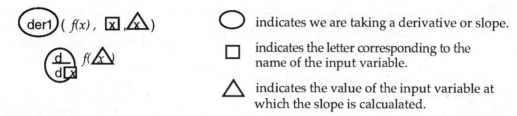

◯ indicates we are taking a derivative or slope.

▢ indicates the letter corresponding to the name of the input variable.

△ indicates the value of the input variable at which the slope is calcualated.

Suppose you are asked to construct a table of values of $f'(x)$ where $f(x) = x^2$ evaluated at different values of x. Two methods of doing this are illustrated below:

Return to the home screen and type the expression on the right. Access der1 with [2nd] [÷] (CALC) [F3].
 	
 der1(X², X, 2) = $\dfrac{dy}{dx}$ for $y = x^2$ evaluated at $x = 2$.
	`der1(x²,x,2)`
	
	`                    4`

Recall the last entry, and edit the expression der1(X², X, 2) by changing the 2 to a ⁻3.
 	
 Press [ENTER]. Continue with this process until you have found all the slope values.
	`                    4`
	`der1(x²,x,⁻3)`
	`                   ⁻6`
	`der1(x²,x,⁻1)`
	`                   ⁻2`
	`der1(x²,x,0)`
	`                    0`

You might prefer to use the TABLE. If so, recall that you must have the expression being evaluated in the y(x)= list. (On the TI-85, it must be in y1.)
 	
 If you enter the slope formula as indicated on the right, you will only have to change y1 when you work with a different function.
	`y1▧x²`
	`y2▧der1(y1,x,x)`
	
	`         x    "   INSF  DELf  SELCT`
	`evalF  nDer  der1  der2  fnInt ▶`

TI-85 Run program TABLE, and choose the AUTO option. Set TblMin = ⁻3 and ΔTbl = 1. Choose to generate 6 table values.	`Have function(s) in`
	`y(x)= list.`
	`Enter 1 (AUTO) or`
	`2 (ASK) 1`
	`TblMin? ⁻3`
	`ΔTbl? 1`
	`Number of values? 6`

Press [ENTER] when the message appears to see the points (x, y1(x)) in the first column and the points (x, der1(y1, x, a) in the second column for a = ⁻3, ⁻2, ⁻1, 0, 1, and 2.	`scroll through table.`
	`  [[(⁻3,9)  (⁻3,⁻6)]`
	`   [(⁻2,4)  (⁻2,⁻4)]`
	`   [(⁻1,1)  (⁻1,⁻2)]`
	`   [(0,0)   (0,0)  ]`
	`   [(1,1)   (1,2)  ]`
	`   [(2,4)   (2,4)  ]]`

TI-86 You can either type in the *x*-values using the ASK feature of the table or you can set TblMin = ⁻3 and ΔTbl = 1 with the AUTO setting chosen.	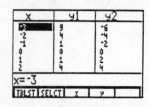
Evaluate the function and the numerical derivative at *x* = ⁻3, ⁻2, ⁻1, 0, 1, 2, and 3. Determine a relationship (pattern) between the slopes and the values of *x*. Function values appear in the y1 column and slopes appear in the y2 column.	

Both Determine a relationship (pattern) between the slopes and the values of *x*. Function values appear in the y1 column and slopes appear in the y2 column.	The TI-85 and the TI-86 only calculate or approximate numerical values of slopes -- they do not give the slope in formula form.

- If you have difficulty determining a pattern, enter the *x*-values at which you are evaluating the slope in list L1 and the values of der1 in list L2. Draw a scatter plot of the *x*-values and the slope values. The shape of the scatter plot should give you a clue as to the equation of the slope formula. If not, try drawing another scatter plot where L1 contains the values of *y* = *f(x)* and L2 contains the calculated slope formula values. Note that this method might help only if you consider a variety of values for *x* in list L1.

4.3.3 DRAWING TANGENT LINES FROM THE GRAPHICS SCREEN Chapter 3

of this *Guide* (specifically, Section 3.3.2) presented a method of drawing tangent lines from the home screen. We now examine another method for drawing tangent lines, this time using the graphics screen. You may not find this method as useful as the previous one, however, because the point at which the tangent line is drawn depends on the horizontal settings in the viewing window.

We illustrate this method of drawing tangent lines with $f(x) = 2\sqrt{x-5}$. Without the context of a real-world situation, how do you know what input values to consider? The answer is that you need to call upon your knowledge of functions. Remember that we graph only real numbers. If the quantity under the square root symbol is negative, the output of *f(x)* is not a real number. We therefore know that *x* must be greater than or equal to 5. Many different horizontal views will do, but we choose to use $0 \le x \le 15$. You can use previously-discussed methods to set height of the window, or you can use the one given below.

Either clear or turn off any previously-entered functions. Enter $f(x) = 2\sqrt{x-5}$ in y1. The parentheses around the *x* – 5 are necessary to include the entire quantity under the square root.	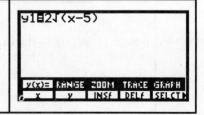

Set the range (window) to that shown on the right. Press F5 (GRAPH) to graph the function.	RANGE xMin=0 xMax=15 xScl=5 yMin=-5 yMax=10 yScl=1 y(x)= RANGE ZOOM TRACE GRAPH▸
TI-85 With the graph on the screen, press MORE F1 (MATH) MORE MORE F3 (TANLN) **TI-86** With the graph on the screen, press MORE F1 (MATH) MORE MORE F1 (TANLN) **Both** Use ▶ or ◀ to move to some point on the curve. Press ENTER.	1 x=7.1428571429 y=2.9277002188
The tangent line is drawn at the position of the cursor. At the bottom of the screen, the calculator also gives the slope of the tangent line at that point.	dy/dx=.68313005106
TI-85 Suppose you want the tangent line drawn at $x = 8$. Press MORE F1 (MATH) MORE MORE F3 (TANLN) and use ▶ or ◀ to move to $x = 8$. You can't do it! With the horizontal settings for this screen, it is not possible to obtain $x = 8$ by tracing.	1 x=8.0952380952 y=3.5186577527
TI-86 With the graph on the screen, press MORE F1 (MATH) MORE MORE F1 (TANLN). The TI-86 allows you to input the exact value of the point at which you want the tangent drawn, even if that value can not be obtained by tracing. Press 8 and ENTER. The line and the slope at $x = 8$ are displayed.	dy/dx=.57735026919

Both If you want to draw the tangent line at a certain value of the input variable that is not a possible trace value, return to the home screen and use the method given in Section 3.3.2 of this *Guide*. (On the TI-86, either method can be used.) It is very important to remember the situations discussed in that section in which the instantaneous rate of change does not exist, but for which the calculator's tangent line incorrectly draws on the screen and the situations in which it doesn't draw but should.

4.3.4 CALCULATING $\frac{dy}{dx}$ AT SPECIFIC INPUT VALUES Section 4.3.1 of this *Guide*

examined two forms of the calculator's numerical derivative and illustrated that one gives a good approximation of the slope of the tangent line and the other gives the slope of the tangent line for most functions at points where the instantaneous rate of change exists. You can also evaluate the calculator's numerical derivative from the graphics screen using the

GRAPH MATH menu. However, instead of being called der1 in that menu, it is called $\frac{dy}{dx}$. We illustrate its use with the function $f(x) = 2\sqrt{x-5}$.

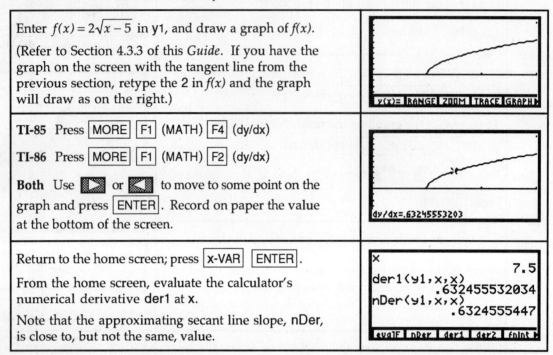

Enter $f(x) = 2\sqrt{x-5}$ in y1, and draw a graph of $f(x)$. (Refer to Section 4.3.3 of this *Guide*. If you have the graph on the screen with the tangent line from the previous section, retype the 2 in $f(x)$ and the graph will draw as on the right.)	
TI-85 Press MORE F1 (MATH) F4 (dy/dx) **TI-86** Press MORE F1 (MATH) F2 (dy/dx) **Both** Use ▶ or ◀ to move to some point on the graph and press ENTER. Record on paper the value at the bottom of the screen.	
Return to the home screen; press x-VAR ENTER. From the home screen, evaluate the calculator's numerical derivative der1 at x. Note that the approximating secant line slope, nDer, is close to, but not the same, value.	

- The value at which you evaluated the calculator's numerical derivative is stored in x. (The values you see probably will not be the same as those displayed on the above screens.)

📖 4.4 The Sum Rule, 4.5 The Chain Rule, and 4.6 The Product Rule

If you have time, it is always a good idea to check your answer. Although your calculator cannot give you a general rule for the derivative of a function, you can use graphical and numerical techniques to check your derivative formula answers. These same procedures apply when you check your results after applying the Sum Rule, the Chain Rule, or the Product Rule.

4.4.1 NUMERICALLY CHECKING SLOPE FORMULAS
When you use a formula to find the derivative of a function, it is possible to check your answer using the calculator's numerical derivative der1. The basic idea of the checking process is that if you evaluate your derivative and the calculator's numerical derivative at several randomly chosen values of the input variable and the output values are very close to the same values, your derivative is *probably* correct.

The average yearly fuel consumption per car in the United States from 1980 through 1990 can be modeled by $g(t) = 0.775t^2 - 140.460t + 6868.818$ gallons per car where t is the number of years since 1900. Applying the sum, power, and constant multiplier rules for derivatives, suppose you determine $g'(t) = 1.55t - 140.460$ gallons per year per car. We now numerically check this answer. (As we have mentioned several times, if you have found a model from data, you should have the complete model, not the rounded one given by $g(t)$, in the y(x)= list of the calculator.)

Enter the function you are taking the derivative of in y1, the calculator's derivative in y2, and your derivative formula, $\frac{dg}{dt} = g'(t)$, in y3. (Remember, since we are going to graph in the y(x)= list, the input variable must be called x.)	

Since the *g(t)* model represents average fuel consumption per car where $t = 80$ in 1980, it makes sense to use only whole number values of $t = x$ greater than or equal to 80 when checking the derivative formula. Turn off y1 by having the cursor on the y1 line and pressing ⬛F5⬛ (SELCT) since you are checking to see if y2 ≈ y3.

| **TI-85** Run program **TABLE** and choose option 2 for **ASK**. Input $x = 80$ and see if the values of y2 and y3 are very close to the same. (Recall that we are using a rounded derivative in y3.)  Repeat the process for at least 4 other values of $x$. If **all** the values of y2 and y3 are very close, your formula for the derivative is *probably* correct. | ```
[(input, output)].
1st column is for y1,
2nd column for y2,...

Press right arrow to
scroll through table.
...-16.45058) (80, -16....
``` |
|---|---|

TI-86 In **TABLE SETUP**, choose **ASK** in the Indpnt location. Press ⬛F1⬛ (TABLE) and check to see that y2 and y3 are very close to the same values for at least four values of x. (Recall that we are using a rounded derivative in y3. This accounts for most of the differences in the two columns of output values.)	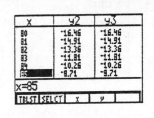

- If the two columns of output values are *not* very close to the same, you have either incorrectly entered a function in the y(x)= list or made a mistake in your derivative formula.

4.4.2 GRAPHICALLY CHECKING SLOPE FORMULAS

Another method of checking your answer for a slope formula (derivative) is to draw the graph of the calculator's numerical derivative and draw the graph of your derivative. If the graphs appear identical *in the same viewing window*, your derivative is probably correct.

We again use the fuel consumption functions from Section 4.4.1 of this *Guide*.

TI-85 Have the function you are taking the derivative of in y1, the calculator's numerical derivative in y2, and your derivative formula, $\frac{dg}{dt} = g'(t)$, in y3.	
Turn off y1 and y2 so that only the graph of y3 will draw. Set an appropriate viewing window such as x between 80 and 90 and y between ⁻25 and 25 or draw the graph with ZFIT in the ZOOM menu.	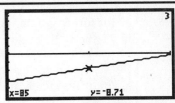
```

Now, turn off y3, turn on y2, and draw the graph of y2 *in the same viewing window.*

(The graph of the calculator's numerical derivative der1 takes slightly longer to draw because the calculator computes the output before plotting each point.)

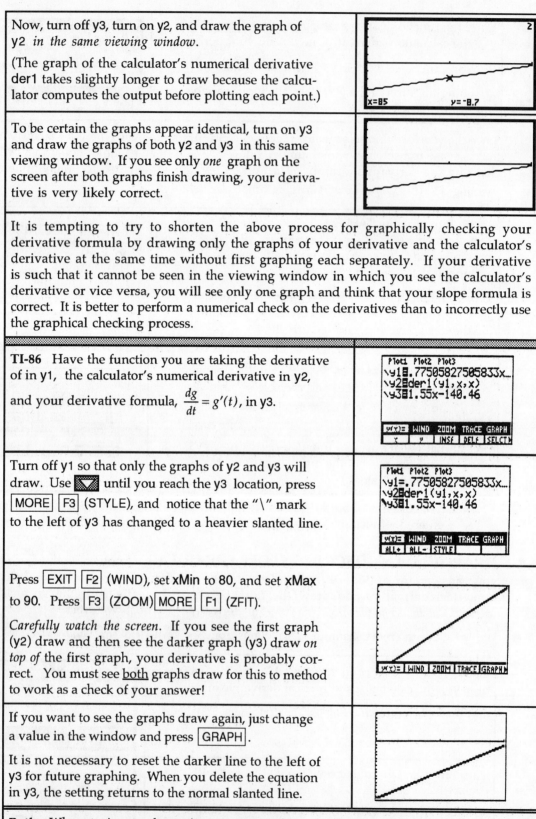

To be certain the graphs appear identical, turn on y3 and draw the graphs of both y2 and y3 in this same viewing window. If you see only *one* graph on the screen after both graphs finish drawing, your derivative is very likely correct.

It is tempting to try to shorten the above process for graphically checking your derivative formula by drawing only the graphs of your derivative and the calculator's derivative at the same time without first graphing each separately. If your derivative is such that it cannot be seen in the viewing window in which you see the calculator's derivative or vice versa, you will see only one graph and think that your slope formula is correct. It is better to perform a numerical check on the derivatives than to incorrectly use the graphical checking process.

TI-86 Have the function you are taking the derivative of in y1, the calculator's numerical derivative in y2, and your derivative formula, $\frac{dg}{dt} = g'(t)$, in y3.

Turn off y1 so that only the graphs of y2 and y3 will draw. Use ▼ until you reach the y3 location, press MORE F3 (STYLE), and notice that the "\" mark to the left of y3 has changed to a heavier slanted line.

Press EXIT F2 (WIND), set xMin to 80, and set xMax to 90. Press F3 (ZOOM) MORE F1 (ZFIT).

Carefully watch the screen. If you see the first graph (y2) draw and then see the darker graph (y3) draw *on top of* the first graph, your derivative is probably correct. You must see <u>both</u> graphs draw for this to method to work as a check of your answer!

If you want to see the graphs draw again, just change a value in the window and press GRAPH .

It is not necessary to reset the darker line to the left of y3 for future graphing. When you delete the equation in y3, the setting returns to the normal slanted line.

Both When trying to determine an appropriate viewing window, read the problem again; it will likely indicate the values for xMin and xMax. Also use your knowledge of the general shape of the function being graphed to know what you should see.

Chapter 5   Analyzing Change: Extrema and Points of Inflection

📖 5.1 Optimization

Your calculator can be very helpful in checking your analytic work when you find optimal points and points of inflection. When you are not required to show work using derivatives or when a very good approximation to the exact answer is all that is required, it is a very simple process to use your calculator to find optimal points and inflection points.

5.1.1 FINDING X-INTERCEPTS OF SLOPE GRAPHS   Where the graph of a function has a local maximum or minimum, the slope graph has a horizontal tangent. Where the tangent line is horizontal, the derivative of the function is zero. Thus, finding where the slope graph *crosses* the input axis is the same as finding where a relative maximum or a relative minimum occurs.

Consider, for example, the model for a cable company's revenue for the 26 weeks after it began a sales campaign:

$$R(x) = {}^-3x^4 + 160x^3 - 3000x^2 + 24{,}000x \quad \text{dollars}$$

where x is the number of weeks since the cable company began sales.

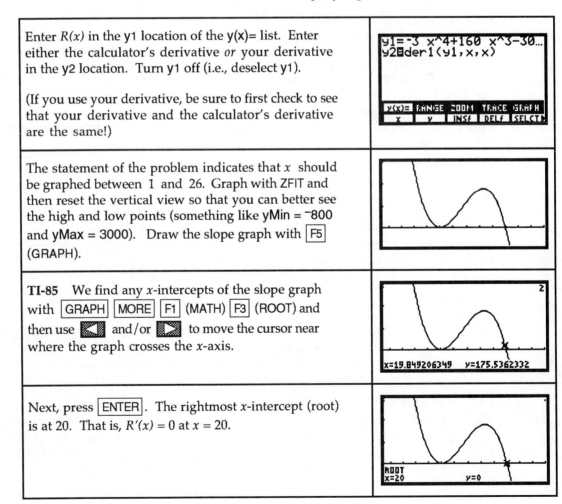

Enter $R(x)$ in the y1 location of the y(x)= list. Enter either the calculator's derivative *or* your derivative in the y2 location. Turn y1 off (i.e., deselect y1). (If you use your derivative, be sure to first check to see that your derivative and the calculator's derivative are the same!)	
The statement of the problem indicates that x should be graphed between 1 and 26. Graph with ZFIT and then reset the vertical view so that you can better see the high and low points (something like yMin = ⁻800 and yMax = 3000). Draw the slope graph with F5 (GRAPH).	
TI-85 We find any x-intercepts of the slope graph with GRAPH MORE F1 (MATH) F3 (ROOT) and then use ◀ and/or ▶ to move the cursor near where the graph crosses the x-axis.	
Next, press ENTER. The rightmost x-intercept (root) is at 20. That is, $R'(x) = 0$ at $x = 20$.	

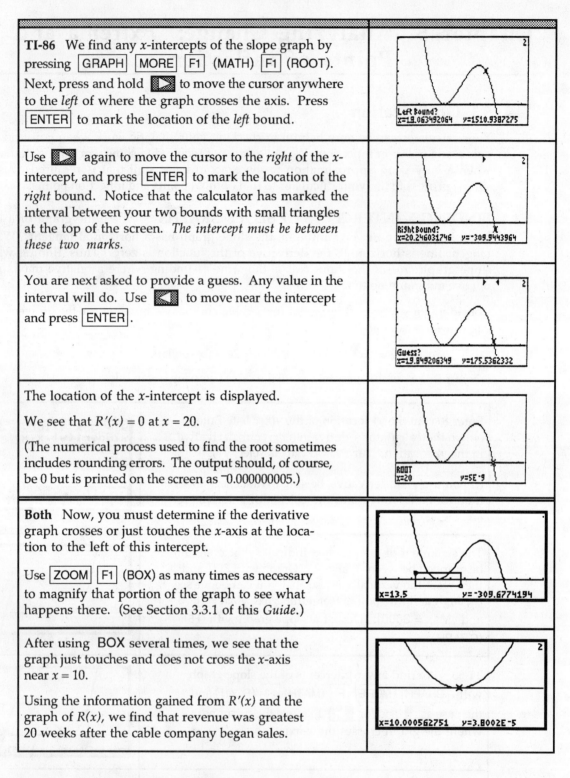

TI-86 We find any *x*-intercepts of the slope graph by pressing GRAPH MORE F1 (MATH) F1 (ROOT). Next, press and hold ▶ to move the cursor anywhere to the *left* of where the graph crosses the axis. Press ENTER to mark the location of the *left* bound.

Use ▶ again to move the cursor to the *right* of the *x*-intercept, and press ENTER to mark the location of the *right* bound. Notice that the calculator has marked the interval between your two bounds with small triangles at the top of the screen. *The intercept must be between these two marks.*

You are next asked to provide a guess. Any value in the interval will do. Use ◀ to move near the intercept and press ENTER.

The location of the *x*-intercept is displayed.

We see that $R'(x) = 0$ at $x = 20$.

(The numerical process used to find the root sometimes includes rounding errors. The output should, of course, be 0 but is printed on the screen as ⁻0.000000005.)

Both Now, you must determine if the derivative graph crosses or just touches the *x*-axis at the location to the left of this intercept.

Use ZOOM F1 (BOX) as many times as necessary to magnify that portion of the graph to see what happens there. (See Section 3.3.1 of this *Guide*.)

After using BOX several times, we see that the graph just touches and does not cross the *x*-axis near $x = 10$.

Using the information gained from $R'(x)$ and the graph of $R(x)$, we find that revenue was greatest 20 weeks after the cable company began sales.

5.1.2 FINDING OPTIMAL POINTS   Once you draw a graph of a function that clearly shows any optimal points, finding the location of those high points and low points is an easy task for your calculator. When a relative maximum or a relative minimum exists at a point, your calculator can find it in a few simple steps. We again use the cable company revenue equation, *R(x)*, from Section 5.1.1.

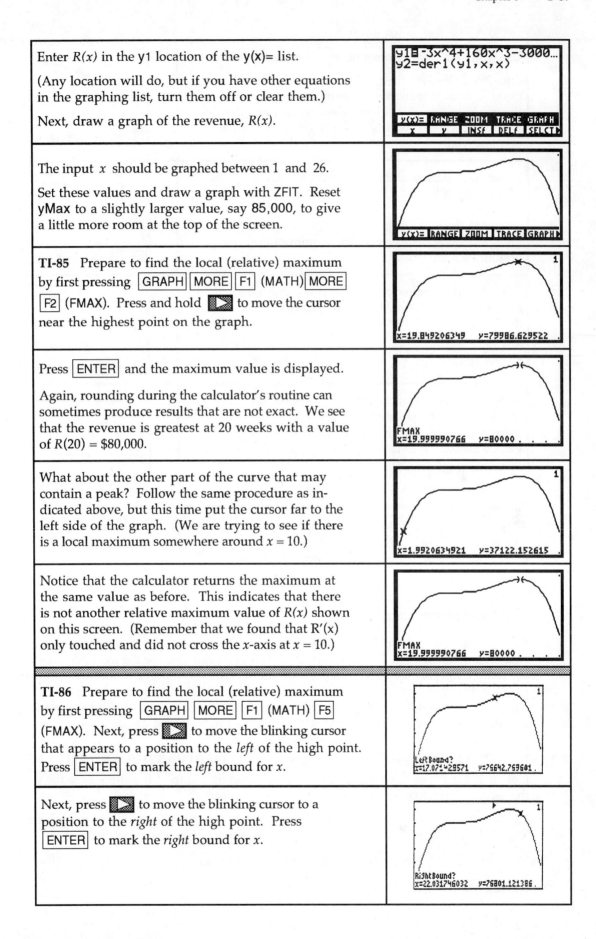

Enter *R(x)* in the y1 location of the y(x)= list.

(Any location will do, but if you have other equations in the graphing list, turn them off or clear them.)

Next, draw a graph of the revenue, *R(x)*.

The input *x* should be graphed between 1 and 26.

Set these values and draw a graph with ZFIT. Reset **yMax** to a slightly larger value, say 85,000, to give a little more room at the top of the screen.

TI-85 Prepare to find the local (relative) maximum by first pressing GRAPH MORE F1 (MATH) MORE F2 (FMAX). Press and hold ▶ to move the cursor near the highest point on the graph.

Press ENTER and the maximum value is displayed.

Again, rounding during the calculator's routine can sometimes produce results that are not exact. We see that the revenue is greatest at 20 weeks with a value of *R*(20) = $80,000.

What about the other part of the curve that may contain a peak? Follow the same procedure as indicated above, but this time put the cursor far to the left side of the graph. (We are trying to see if there is a local maximum somewhere around *x* = 10.)

Notice that the calculator returns the maximum at the same value as before. This indicates that there is not another relative maximum value of *R(x)* shown on this screen. (Remember that we found that R'(x) only touched and did not cross the *x*-axis at *x* = 10.)

TI-86 Prepare to find the local (relative) maximum by first pressing GRAPH MORE F1 (MATH) F5 (FMAX). Next, press ▶ to move the blinking cursor that appears to a position to the *left* of the high point. Press ENTER to mark the *left* bound for *x*.

Next, press ▶ to move the blinking cursor to a position to the *right* of the high point. Press ENTER to mark the *right* bound for *x*.

Use ◀ to move the cursor near your estimate of the high point and press ENTER. The TI-86 uses your guess to locate the highest point in the region between the two bound marks. The maximum value and the x-value at which it occurs is displayed at the bottom of the screen.	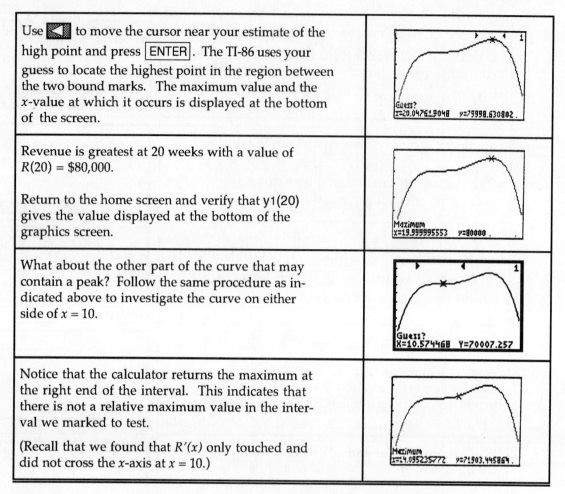
Revenue is greatest at 20 weeks with a value of $R(20) = \$80,000$. Return to the home screen and verify that y1(20) gives the value displayed at the bottom of the graphics screen.	
What about the other part of the curve that may contain a peak? Follow the same procedure as indicated above to investigate the curve on either side of $x = 10$.	
Notice that the calculator returns the maximum at the right end of the interval. This indicates that there is not a relative maximum value in the interval we marked to test. (Recall that we found that $R'(x)$ only touched and did not cross the x-axis at $x = 10$.)	

Both   The methods of this section also apply to finding relative or local minimum values of a function. The only difference is that to find the minimum instead of the maximum, you would use the FMIN menu key that is to the left of the FMAX menu key.

📖 5.2 Inflection Points

As was the case with optimal points, your calculator can be very helpful in checking your analytic work when you find points of inflection. You can also use the methods illustrated in Section 5.1.2 of this *Guide* to find the location of any maximum or minimum points on the graph of the first derivative to find the location of any inflection points for the function. In fact, your calculator offers three graphical methods for finding inflection points. We investigate these as well as the analytic method in the following discussions.

5.2.1   FINDING X-INTERCEPTS OF A SECOND DERIVATIVE GRAPH   We first

look at using the analytic method of finding inflection points -- finding where the graph of the second derivative of a function *crosses* the input axis.

To illustrate, consider a model for the percentage of students graduating from high school in South Carolina from 1982 through 1990 who entered post-secondary institutions:

$$f(x) = {}^{-}0.1057x^3 + 1.355x^2 - 3.672x + 50.792 \text{ percent}$$

where $x = 0$ in 1982.

Enter $f(x)$ in the y1 location of the y(x)= list, your formula for the first derivative in y2 and your second derivative of the function in y3.
 	
Turn off y1 and y2.
	`y1=-.1057x^3+1.355x²...`
	`y2=3(-.1057)x²+2(1.3...`
	`y3⬛6(-.1057x)+2(1.35...`
	
	`Y(X)= RANGE ZOOM TRACE GRAPH`
	`  X  │  Y  │ INSf │ DELf │SELCT▶`

The problem says the model is for 1982 through 1990 which corresponds to $0 \le x \le 8$. Thus, you are told the horizontal view. Choose an appropriate vertical view -- possibly y between $^-4$ and 4. Graph $f''(x)$ Remember that you should be able to clearly see any optimal points. Leave room at the bottom of the screen so that trace coordinates will not block your view of any important points on the graph.	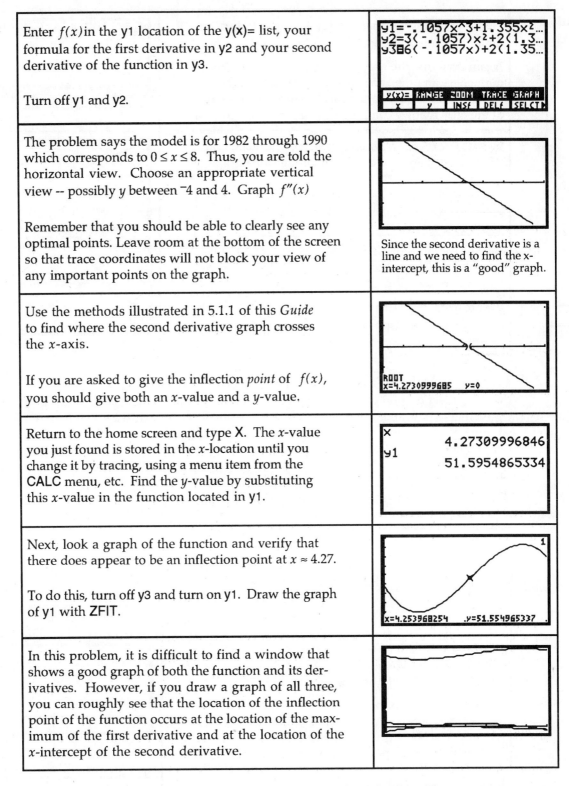 Since the second derivative is a line and we need to find the x-intercept, this is a "good" graph.
Use the methods illustrated in 5.1.1 of this *Guide* to find where the second derivative graph crosses the x-axis. If you are asked to give the inflection *point* of $f(x)$, you should give both an x-value and a y-value.	`ROOT` `x=4.2730999685   y=0`

Return to the home screen and type X. The x-value you just found is stored in the x-location until you change it by tracing, using a menu item from the CALC menu, etc. Find the y-value by substituting this x-value in the function located in y1.	`x`
	`        4.27309996846`
	`y1`
	`       51.5954865334`

Next, look a graph of the function and verify that there does appear to be an inflection point at $x \approx 4.27$. To do this, turn off y3 and turn on y1. Draw the graph of y1 with ZFIT.	`x=4.253968254   .y=51.554965337`
In this problem, it is difficult to find a window that shows a good graph of both the function and its derivatives. However, if you draw a graph of all three, you can roughly see that the location of the inflection point of the function occurs at the location of the maximum of the first derivative and at the location of the x-intercept of the second derivative.	

Both the TI-85 and the TI-86 have a second derivative function. The calculator notation for $f''(x)$, the second derivative, is der2 and it is accessed with 2nd ÷ (CALC) F4 (der2). Use der2 either to check your second derivative formula or enter it in the graphing list instead of your second derivative formula in the process described above.

Enter $f(x)$ in the y1 location of the y(x)= list, der1 in the y2 location, and der2 in the y3 location. Turn off y1. Again note that the only difference in what we are doing here and the above discussion is that we are using the calculator's numerical derivatives instead of the ones that we computed using derivative formulas.	y1=-.1057x^3+1.355x²... y2█der1(y1,x,x) y3█der2(y1,x,x) y(x)█ RANGE ZOOM TRACE GRAPH x y INSf DELf SELCT▶
Be sure $0 \le x \le 8$. Draw the graphs of the slope graph, ($f'(x)$ in y2) and the derivative of the slope graph ($f''(x)$ in y3) using ZFIT.	*graph*
Use the methods of 5.1.1 of this *Guide* to find the x-intercept of the second derivative. (It takes a little longer here than it did before!) Be sure to use ▼ to move to the graph of the line before tracing to the approximate location of the root.	*graph* ROOT X=4.2730999685 y=0

5.2.2 FINDING INFLECTION POINTS WITH YOUR CALCULATOR   Remember

that an inflection point is a point of greatest or least slope. Whenever finding the second derivative of a function is tedious or you do not need an exact answer from an analytic solution, you can very easily find an inflection point of a function by finding where the first derivative of the function has a maximum or minimum value.

We illustrate this process with the function giving the number of polio cases in 1949:

$$y = \frac{42183.911}{1 + 21484.253e^{-1.248911t}}$$ where $t = 1$ on January 31, 1949, $t = 2$ on February 28, 1949, etc.

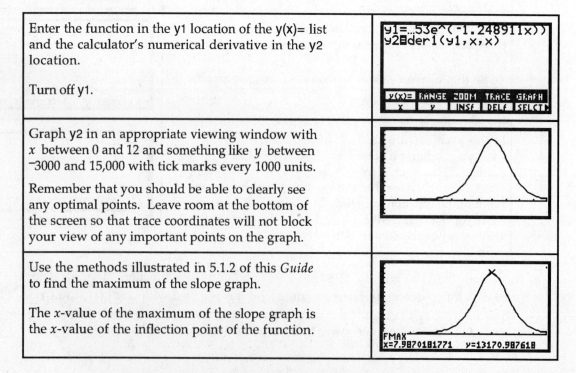

Enter the function in the y1 location of the y(x)= list and the calculator's numerical derivative in the y2 location. Turn off y1.	y1=...53e^(-1.248911x)) y2█der1(y1,x,x) y(x)█ RANGE ZOOM TRACE GRAPH x y INSf DELf SELCT▶
Graph y2 in an appropriate viewing window with x between 0 and 12 and something like y between ‾3000 and 15,000 with tick marks every 1000 units. Remember that you should be able to clearly see any optimal points. Leave room at the bottom of the screen so that trace coordinates will not block your view of any important points on the graph.	*graph*
Use the methods illustrated in 5.1.2 of this *Guide* to find the maximum of the slope graph. The x-value of the maximum of the slope graph is the x-value of the inflection point of the function.	*graph* FMAX X=7.9870181771 y=13170.987618

If you are asked to give the inflection point, you should give both an *x*-value and a *y*-value. Find the *y*-value by substituting this *x*-value in the function located in y1.
 	
The rate of change at that time is obtained by substituting this *x*-value in y2.
	x 7.98701817706
	y1
	      21091.9484486
	y2
	      13170.9876177

Our final method is the simplest -- just be certain if you use it that the function does have an inflection point at the location indicated by the calculator.

Draw the graph of the logistic function in y1, *not* its derivative. (That is, turn off y2 and turn on y1.) Have $0 \leq x \leq 12$ and use ZFIT to set the vertical view.	
TI-85 Press GRAPH MORE F1 (MATH) MORE F3 (INFLC), and move the cursor to the approximate location of where the function changes concavity. Press ENTER to find the inflection point.	 INFLC X=7.9870187124 y=21091.9555
TI-86 Press GRAPH MORE F1 (MATH) MORE F1 (INFLC). Next, press ◄ to move the cursor to a position to the *left* of where the function changes concavity. Press ENTER to mark the *left* bound.	 LeftBound? X=7.3333333333 y=12930.611728
Use ► to move the cursor to the *right* of where the function changes concavity. Press ENTER to mark the *right* bound. At the **Guess?** prompt, move the cursor to the approximate location of the inflection point. Press ENTER.	 Guess? X=8.0952380952 y=22515.14583
The inflection point is displayed and marked on the graph.	 INFLC X=7.9870187124 y=21091.9555

Chapter 6   Accumulating Change: Limits of Sums and the Definite Integral

📖 6.1 Results of Change

We have thus far seen how to use the calculator to work with rates of change. In this chapter we consider the results of change. Your calculator has many useful features that will assist you in your study of the accumulation of change.

6.1.1 APPROXIMATIONS WITH LEFT RECTANGLES   The calculator's lists can be used to perform the calculations needed to approximate, using left rectangles, the area between the horizontal axis, a (non-negative) rate of change function, and two input values.

Consider, for example, a model for the number of customers per minute who came to a Saturday sale at a large department store between 9 a.m. and 9 p.m.:

$$c(m) = (4.58904*10^{-8})\, m^3 - (7.78127*10^{-5})\, m^2 + 0.03303\, m + 0.88763$$

customers per minute where m is the number of minutes after 9 a.m.

Enter this model in the y1 location of the y(x)= list. (If you have other equations in the graphing list, clear them.) (Remember that "10 to a power" is denoted by **E** on the calculator. Access **E** with ⎡EE⎤ .)	`y1=4.589036E-8x^3-7.…` y(X)= RANGE ZOOM TRACE GRAPH x y INSf DELf SELCT⟩
Suppose we want to estimate the total number of customers who came to the sale between $x = 0$ and $x = 660$ (12 hours) with 12 rectangles and $\Delta x = 60$. Enter these x-values in list L1. Recall that a quick way to do this is with ⎡2nd⎤ ⎡–⎤ (LIST) ⎡F5⎤ (OPS) ⎡MORE⎤ ⎡F3⎤ (seq) ⎡x-VAR⎤ ⎡,⎤ ⎡x-VAR⎤ ⎡,⎤ 0 ⎡,⎤ 660 ⎡,⎤ 60 ⎡)⎤ ⎡ENTER⎤ . Store the values in list L1 with ⎡STO▸⎤ L ⎡ALPHA⎤ 1. (See Section 2.1.1 of this *Guide*.)	`seq(x,x,0,660,60)` `{0 60 120 180 240 30…` `Ans→L1` `{0 60 120 180 240 30…` ⟨ ⟩ NAMES EDIT OPS Notice that 9 p.m. is 720 minutes after 9 a.m. When using *left*-rectangle areas, the *rightmost* data point is not included.
TI-85 Enter y-values calculated from the model in list L2 using program MDVAL (model values). This program is in the TI-85/86 Appendix. You can have the model entered in any of the first four locations of the y(x)= list. Tell the program which y location at the Function location? prompt. L2 now contains the *heights* of the 12 rectangles.	`MDVAL` `Function location? 1` `{.88763 2.5993967021…`
Since the *width* of each rectangle is 60, the area of each rectangle is 60*height. Enter the *areas* of the 12 rectangles in list L3. Press and hold down ▶ to view all the values in L3 or use the LIST EDIT menu to view them.	`60 L2→L3` `{53.2578 155.9638021…` ⟨ ⟩ NAMES EDIT OPS L1 L2 L3 L4 L5

TI-86 Enter $c(x)$-values calculated from the model in list L2 with the keystrokes [2nd] [+] (STAT) [F2] (EDIT), press ▲ (to darken the name L2), and type y1(L1) with y1 [( ] [ALPHA] L 1 [ )] [ENTER] . List L2 contains the *heights* of the 12 rectangles.	L1 L2 L3 2 0 .88763 60 2.599217 120 3.810026 180 4.579531 240 4.967207 300 5.032528 L2 =y1(L1) 〈 〉 NAMES " OPS ▸
Since the width of each rectangle is 60, the area of each rectangle is 60*height. Enter the *areas* of the 12 rectangles in list L3 by using ▲ to darken the name of list L3 and then typing 60L2 with 6 0 [ALPHA] L 2 [ENTER] .	L1 L2 L3 3 0 .88763 53.2578 60 2.599217 155.953 120 3.810026 228.6015 180 4.579531 274.7719 240 4.967207 298.0324 300 5.032528 301.9517 L3 =60L2 〈 〉 NAMES " OPS ▸
Both On the home screen, find the sum of the areas of the rectangles with [2nd] [−] (LIST) [F5] (OPS) [MORE] [F1] (SUM) [EXIT] [F3] (NAMES) [F3] (L3) [ENTER] . We estimate, using 12 left rectangles, that about 2,573 customers came to the Saturday sale.	sum L3 2573.090369

Note: The values in Table 6.2 in your text and the final result differ slightly than those in your lists. This is because the unrounded model found with the data was used for computations in the text. If you have the unrounded model available, you should use it instead of the rounded model $c(m)$ that is given above.

6.1.2 APPROXIMATIONS WITH RIGHT RECTANGLES

When using left rectangles to approximate the results of change, the rightmost data point is not the height of a rectangle and is not used in the computation of the left-rectangle area. Similarly, when using right rectangles to approximate the results of change, the *leftmost* data point is not the height of a rectangle and is not used in the computation of the right-rectangle area.

The following data shows the rate of change of the concentration of a drug in the blood stream in terms of the number of days since the drug was administered:

Day	1	5	9	13	17	21	25	29
Concentration ROC (μg/mL/day)	1.5	0.75	0.33	0.20	0.10	‾1.1	‾0.60	‾0.15

First, we fit a piecewise model to the data.

Clear all lists. Enter the days in L1 and the rate of change of concentration in L2. Draw a scatter plot of the data. It is obvious that a piecewise model should be used with $x = 20$ as the "break" point.	
TI-85 Go to the list editor and delete the last three data points from lists L1 and L2 with [F2] (DELi). Enter the list {21, 25, 29} in L3 and in L4 enter the corresponding outputs: {‾1.1, ‾0.60, ‾0.15}. Fit an exponential model to the data in L1 and L2 and put that model in y1.	〈1 5 9 13 17〉 L2 〈1.5 .75 .33 .2 .1〉 L3 〈21 25 29〉 L4 〈-1.1 -.6 -.15〉

Fit a linear model to the data in L3 and L4. When you go to the $\boxed{\text{STAT}}$ (CALC) menu, you must tell the calculator that you are using different lists by entering L3 as the xlist Name and L4 as the ylist Name.

```
xlist Name=L3
ylist Name=L4




 CALC  EDIT  DRAW  FCST
  L4    L5    L6    L8    U ▸
```

Then, fit a linear model with $\boxed{\text{ENTER}}$ $\boxed{\text{F2}}$ (LINR).

Copy the linear model to location y2.

```
LinR
 a=-3.58541666667
 b=.11875
 corr=.999538638961
 n=3
Name=y2
 CALC  EDIT  DRAW  FCST
P2REG P3REG P4REG STREG
```

TI-86 Go to the lists with $\boxed{\text{2nd}}$ $\boxed{+}$ (STAT) $\boxed{\text{F2}}$ (EDIT). Delete the last three data points from lists L1 and L2 by going to the location of those values and pressing $\boxed{\text{DEL}}$. Enter {21, 25, 29} in L3 and in L4 enter the corresponding outputs: {⁻1.1, ⁻0.60, ⁻0.15}.

Fit an exponential model to the data in L1 and L2 and put that model in y1.

```
L2      L3      L4      4
1.5     21     -1.1
.75     25     -.6
.33     29     -.15
.2      -----
.1
-------
L4(4) =
   {    }   NAMES   "    OPS ▸
```

Fit a linear model to the data in lists L3 and in L4 and copy the model to location y2 with $\boxed{\text{2nd}}$ $\boxed{+}$ (STAT) $\boxed{\text{F1}}$ (CALC) $\boxed{\text{F3}}$ (LinR) L3 $\boxed{,}$ L4 $\boxed{,}$ y2

```
LinR L3,L4,y2





 CALC  EDIT  PLOT  DRAW  VARS
OneVa TwoVa LinR  LnR  ExpR ▸
```

Both (optional) If you want to graph this model on the original scatter plot, first re-enter the data with all the inputs in list L1, all the outputs in list L2.

Return to the graphing list and turn *off* y1 and y2.

Enter (y1)(x ≤ 20) + (y2)(x > 20) in y3. Choose DrawDot format, draw the scatter plot and y3 will graph over it.

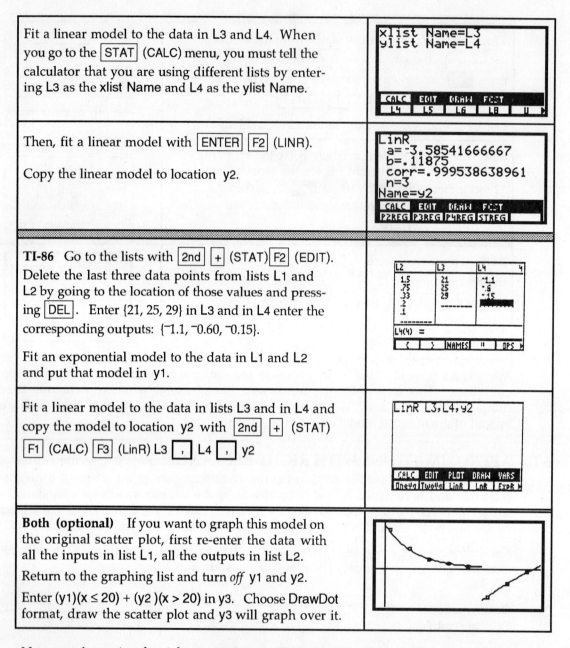

Now, we determine the right-rectangle area for $0 \leq x \leq 20$:

Clear all lists. Because x starts at 0 and $\Delta x = 2$, enter in L1 the values 0, 2, 4, 6, …, 20 or use seq (x, x, 0, 20, 2) to generate the list.

Then, because we are using right rectangles, go to the list editor and *delete* the leftmost input value (0).

```
seq(x,x,0,20,2)
{0 2 4 6 8 10 12 14 …
Ans→L1
{0 2 4 6 8 10 12 14 …
L1
{2 4 6 8 10 12 14 16…
```

TI-85 Enter y-values calculated from the model in y1 into list L2 using program MDVAL.

L2 now contains the *heights* of the 12 rectangles.

```
MDVAL
Function location? 1
{1.21962608434 .8707…
L2
{1.21962608434 .8707…
```

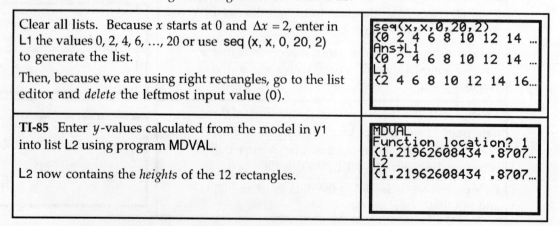

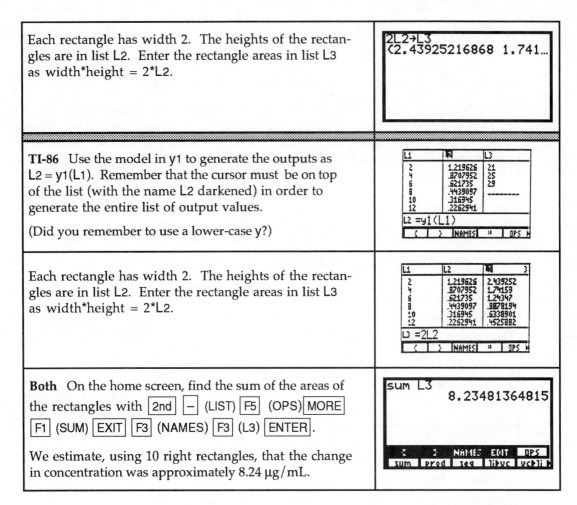

Each rectangle has width 2. The heights of the rectangles are in list L2. Enter the rectangle areas in list L3 as width*height = 2*L2.

TI-86 Use the model in y1 to generate the outputs as L2 = y1(L1). Remember that the cursor must be on top of the list (with the name L2 darkened) in order to generate the entire list of output values.

(Did you remember to use a lower-case y?)

Each rectangle has width 2. The heights of the rectangles are in list L2. Enter the rectangle areas in list L3 as width*height = 2*L2.

Both On the home screen, find the sum of the areas of the rectangles with [2nd] [−] (LIST) [F5] (OPS) [MORE] [F1] (SUM) [EXIT] [F3] (NAMES) [F3] (L3) [ENTER].

We estimate, using 10 right rectangles, that the change in concentration was approximately 8.24 µg/mL.

To estimate, using right rectangles, the change in drug concentration for $20 \le x \le 29$ days with $\Delta x = 1$, follow the same procedure as above. The values in L1 begin with 21 because we must eliminate the leftmost value (20) when using right rectangles. You should use the absolute value of the model in y2 to generate L2.

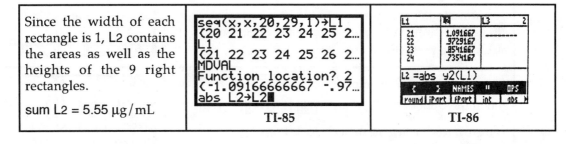

Since the width of each rectangle is 1, L2 contains the areas as well as the heights of the 9 right rectangles.

sum L2 = 5.55 µg/mL

6.2 Trapezoid and Midpoint Rectangle Approximations

You can compute areas of trapezoids on the home screen of your calculator or use the fact that the trapezoid approximation is the average of the left- and right-rectangle approximations. Areas of midpoint rectangles are found in the same manner as left and right rectangle areas except that the midpoint of the base of each rectangle is in L1 and no data values are deleted. However, such procedures can become tedious when the number, n, of subintervals is large.

6.2.1 SIMPLIFYING AREA APPROXIMATIONS

When you have a model $y = f(x)$ in location y1 in the y(x)= list, you will find program NUMINTGL very helpful in determining left-rectangle, right-rectangle, midpoint-rectangle, and trapezoidal numerical approximations for accumulated change. Program NUMINTGL is listed in the TI-85/TI-86 Appendix.

WARNING: This program will not work properly if you have any functions in your y(x)= list that have letters other than x in them. Delete any such functions (like the logistic equation) before using program NUMINTGL. If you receive an error while running the program, you may have a picture or program stored to a single-letter name. For instance, if program NUMINTGL is trying to store a number in T and you have a program called T, the calculator stops. Delete or rename any programs or pictures that you have called by a single-letter name before continuing.

We illustrate using this program with a model for the Carson River flow rates:

$$f(h) = 18{,}225h^2 - 135{,}334.3h + 2{,}881{,}542.9 \text{ cubic feet per hour}$$

h hours after 11:45 a.m. Wednesday. (The complete model found from the data in your text is used for all the following calculations.)

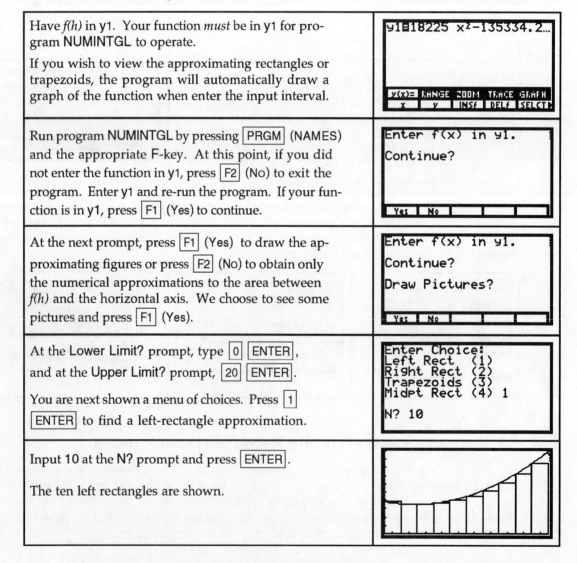

Have $f(h)$ in y1. Your function *must* be in y1 for program NUMINTGL to operate. If you wish to view the approximating rectangles or trapezoids, the program will automatically draw a graph of the function when enter the input interval.	`y1◻18225 x²-135334.2…`
Run program NUMINTGL by pressing PRGM (NAMES) and the appropriate F-key. At this point, if you did not enter the function in y1, press F2 (No) to exit the program. Enter y1 and re-run the program. If your function is in y1, press F1 (Yes) to continue.	`Enter f(x) in y1.` `Continue?`

At the next prompt, press F1 (Yes) to draw the approximating figures or press F2 (No) to obtain only the numerical approximations to the area between $f(h)$ and the horizontal axis. We choose to see some pictures and press F1 (Yes).	`Enter f(x) in y1.`
	`Continue?`
	`Draw Pictures?`

At the **Lower Limit?** prompt, type 0 ENTER, and at the **Upper Limit?** prompt, 20 ENTER.
 	
You are next shown a menu of choices. Press 1 ENTER to find a left-rectangle approximation.
	`Enter Choice:`
	`Left Rect  (1)`
	`Right Rect (2)`
	`Trapezoids (3)`
	`Midpt Rect (4) 1`
	`N? 10`

Input 10 at the N? prompt and press ENTER. The ten left rectangles are shown.	

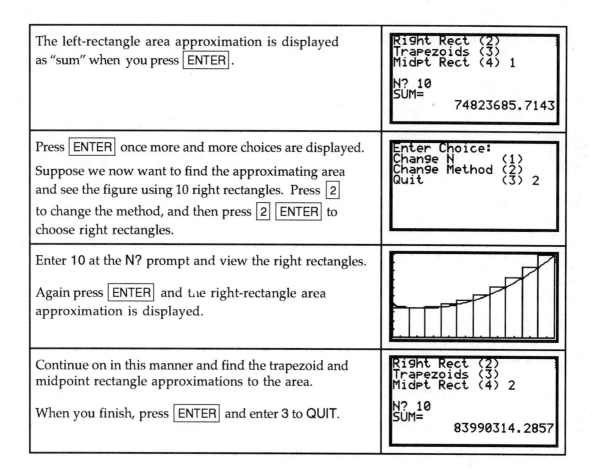

The left-rectangle area approximation is displayed as "sum" when you press ENTER.	``` Right Rect (2) Trapezoids (3) Midpt Rect (4) 1 N? 10 SUM= 74823685.7143 ```
Press ENTER once more and more choices are displayed. Suppose we now want to find the approximating area and see the figure using 10 right rectangles. Press 2 to change the method, and then press 2 ENTER to choose right rectangles.	``` Enter Choice: Change N (1) Change Method (2) Quit (3) 2 ```
Enter 10 at the N? prompt and view the right rectangles. Again press ENTER and the right-rectangle area approximation is displayed.	
Continue on in this manner and find the trapezoid and midpoint rectangle approximations to the area. When you finish, press ENTER and enter 3 to QUIT.	``` Right Rect (2) Trapezoids (3) Midpt Rect (4) 2 N? 10 SUM= 83990314.2857 ```

6.3 The Definite Integral as a Limit of Sums

This section introduces you to a very important and useful concept of calculus -- the definite integral. Your calculator can be very helpful as you study definite integrals and how they relate to the accumulation of change.

6.3.1 LIMITS OF SUMS
When you are looking for a trend in midpoint-rectangle approximations to the area between a non-negative function and the horizontal axis between two values of the input variable, program NUMINTGL is extremely useful! However, for this use of the program, it is not advisable to draw pictures when n, the number of subintervals, is large.

To construct a chart of midpoint approximations for the area between $f(x) = \sqrt{1 - x^2}$ and the x-axis from $x=0$ and $x=1$, first enter $f(x)$ in y1. (Don't forget to enclose $1 - x^2$ in parentheses.)	``` y1=√(1-x²) Y(X)= RANGE ZOOM TRACE GRAPH x y INSf DELf SELCT▶ ```
Start program NUMINTGL. Since many subintervals are used with the idea of a limit of sums, do *not* choose to draw the pictures unless you have time to sit and wait for all the rectangles to draw! After entering 0 and 1 as the respective lower and upper limits, choose option 4 for midpoint rectangles.	``` Enter Choice: Left Rect (1) Right Rect (2) Trapezoids (3) Midpt Rect (4) 4 N? 4 ```

Input some number of subintervals, say N = 4. Record on paper the midpoint area approximation 0.7959823052. (If you want an answer accurate to the thousandths position, record at least 4 decimal places.)	``` Right Rect (2) Trapezoids (3) Midpt Rect (4) 4 N? 4 SUM= .795982305153 ```
Press ENTER and choose option 1: CHANGE N. Double the number of subintervals to N = 8. Record the midpoint approximation 0.7891717328 (again, to at least 4 decimal places).	``` Change N (1) Change Method (2) Quit (3) 1 N? 8 SUM= .789171732825 ```
Continue on in this manner, each time choosing option 1: CHANGE N and doubling N until a trend is evident. (Finding a trend means that you can tell what value the approximations are getting closer and closer to, within a specified accuracy, without having to run the program ad infinitum!) Choose QUIT (3) to stop.	``` Change N (1) Change Method (2) Quit (3) 1 N? 256 SUM= .78541918062 ```

- Remember that the trend indicated by the limit of sums can be interpreted as the area of the region between the function and the input axis *only* when that region lies above the input axis. When the region lies below the input axis, the trend is the negative of the area of the region.

6.5 The Fundamental Theorem

Recall that der1 is the calculator's numerical derivative and provides, in most cases, a good value for the instantaneous rate of change of a function when that rate of change exists. Your calculator can also give you a numerical approximation for a definite integral of a function. This numerical integrator is called fnInt, and the correspondence between the mathematical definite integral notation $\int_a^b f(x)\,dx$ and the calculator's notation fnInt(f(X), X, a, b) is as shown below:

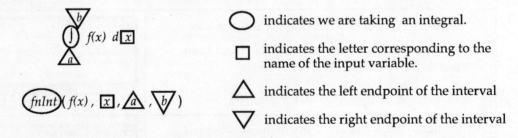

◯ indicates we are taking an integral.

▢ indicates the letter corresponding to the name of the input variable.

△ indicates the left endpoint of the interval

▽ indicates the right endpoint of the interval

You will find that in most cases, your calculator's numerical integrator gives a very close decimal approximation of the exact value of a definite integral. This use is examined in Section 6.6 of this *Guide*. In this section, we illustrate the Fundamental Theorem of Calculus and see how to draw the graph of a general antiderivative of a function.

6.5.1 THE FUNDAMENTAL THEOREM OF CALCULUS

This theorem tells us that the derivative of an antiderivative of a function is the function itself. Let us view this theorem both numerically and graphically. Your calculator's numerical integrator is accessed with the keystrokes 2nd ÷ (CALC) F5 (fnInt(). The fnInt symbol needs to be followed by a function, the variable, a lower limit, and an upper limit (in that order).

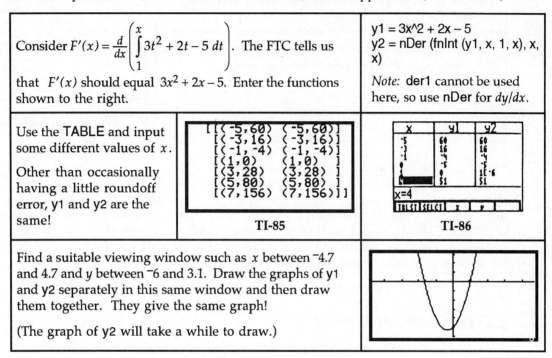

Consider $F'(x) = \dfrac{d}{dx}\left(\displaystyle\int_1^x 3t^2 + 2t - 5\, dt\right)$. The FTC tells us that $F'(x)$ should equal $3x^2 + 2x - 5$. Enter the functions shown to the right.

y1 = 3x^2 + 2x − 5
y2 = nDer (fnInt (y1, x, 1, x), x, x)

Note: der1 cannot be used here, so use nDer for *dy/dx*.

Use the **TABLE** and input some different values of x.

Other than occasionally having a little roundoff error, y1 and y2 are the same!

```
[[(-5,60)   (-5,60)]
 [(-3,16)   (-3,16)]
 [(-1,-4)   (-1,-4)]
 [(1,0)     (1,0)]
 [(3,28)    (3,28) ]
 [(5,80)    (5,80) ]
 [(7,156)   (7,156)]]
```
TI-85

TI-86

Find a suitable viewing window such as x between ⁻4.7 and 4.7 and y between ⁻6 and 3.1. Draw the graphs of y1 and y2 separately in this same window and then draw them together. They give the same graph!

(The graph of y2 will take a while to draw.)

Enter several other functions in y1 and perform the same explorations as above. Confirm your results with derivative and integral formulas. Are you convinced?

6.5.2 DRAWING ANTIDERIVATIVE GRAPHS

All antiderivatives of a specific function differ only by a constant. We explore this idea using the function $f(x) = 3x^2 - 1$ and its antiderivative $F(x) = x^3 - x + C$.

The correct syntax for the calculator's numerical integrator is fnInt(f(X), X, a, b) where $f(x)$ is the function you are integrating, x is the variable of integration, a is the lower limit on the integral, and b is the upper limit on the integral. You do *not* have to use x as the variable unless you are graphing the integral or evaluating it using the calculator's table.

Note that you must supply both a lower limit and an upper limit for fnInt. We can use x for the upper limit, but not both the upper and lower limits. Since we are working with a general antiderivative in this illustration, we do not have the starting point for the accumulation. We therefore just choose some value, say 0, to use as the lower limit to illustrate drawing antiderivative graphs. If you choose a different lower limit, your results will differ from those shown below by a constant.

Enter $f(x)$ in y1, fnInt(y1, x, 0, x) in y2, and $F(x)$ in y3, y4, y5, y6, and y7 (using a different value of C in each location)

(You can try different values of C than those shown on the right.)

```
y1B3x²−1
y2BfnInt(y1,x,0,x)
y3Bx^3−x+0
y4Bx^3−x−2
y5Bx^3−x−1
y6Bx^3−x+2
y7Bx^3−x−3
```
y(x)= | RANGE | ZOOM | TRACE | GRAPH▸

Find a suitable viewing window and graph all the functions. (Try x between ⁻3 and 3 and y between ⁻5 and 10.) It seems that the only difference in the graphs (other than the graph of y1) is that the y-intercept is different. But, isn't C the y-intercept?	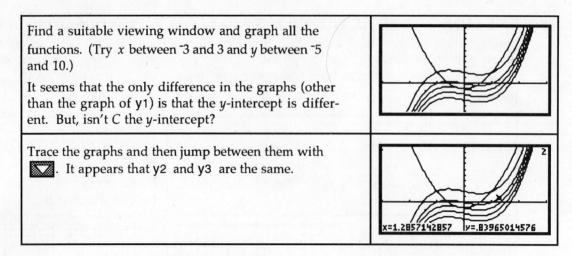
Trace the graphs and then jump between them with ▼. It appears that y2 and y3 are the same.	

Warning: The methods of Sections 4.4.1 and/or 4.4.2 that we used to check antiderivative formulas are not valid with general antiderivatives. Why? The answer is because in this situation you must choose arbitrarily choose a value for the constant of integration and you must arbitrarily choose a value for the lower limit in order to use the calculator's numerical integrator. However, for most rate-of-change functions where $f(0) = 0$, the calculator's numerical integrator values and your antiderivative formula values should differ by the same constant at every input value where they are defined.

📖 6.6 The Definite Integral

When using the numerical integrator on the home screen, enter fnInt($f(x)$, x, a, b) for a specific function $f(x)$ with input x and specific values of a and b. (Remember that the input variable does not have to be x on the home screen.) If you prefer, $f(x)$ can be in the y(x)= list and referred to as y1 (or whatever location you have it in) when using fnInt.

6.6.1 EVALUATING A DEFINITE INTEGRAL ON THE HOME SCREEN

We illustrate the use of your calculator's numerical integrator with the model for the rate of change of the average sea level in meters per year during the last 7000 years. A model for these data is $r(t) = 0.14762t^2 + 0.35952t - 0.8$ meters per year t thousand years from the present. (Note that t is negative since we are talking about the past.)

As we have previously mentioned, you should always use the full model (not rounded) for any model you find from data. Find a quadratic to fit the data shown in the *Changing Sea Levels* example in your text, and enter it in some location of the y(x)= list, say y1.	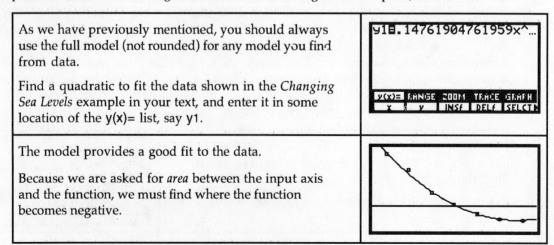
The model provides a good fit to the data. Because we are asked for *area* between the input axis and the function, we must find where the function becomes negative.	

Use the ideas of Section 1.2.3 of this *Guide* to find the x-intercept (root) of y1 between 0 and 7 to be $x \approx {}^-3.845$.	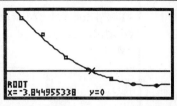
Return to the home screen and type the expression shown to the right. Remember that fnInt is accessed with the keystrokes 2nd ÷ (CALC) F5 (fnInt()). $$\int_{-7}^{-3.845} r(t)\, dt \approx 5.4 \text{ meters}$$	fnInt(y1,x,-7,-3.845) 5.40593487522 evalF nDer der1 der2 fnInt▶

The area of the region above the x-axis is about 5.4 meters. Now, find the area of the region below the x-axis. This area equals the negative of the definite integral of the function over that region.

Evaluate $\int_{-3.845}^{0} r(t)\, dt$ as shown on the right. The area of the region below the x-axis is about 2.9 meters.	fnInt(y1,x,-7,-3.845) 5.40593487522 fnInt(y1,x,-3.845,0) -2.93649043074 evalF nDer der1 der2 fnInt▶
Find $\int_{-7}^{0} r(t)\, dt$. Note that this value is *not* the sum of the two areas. It is their difference.	5.40593487522 fnInt(y1,x,-3.845,0) -2.93649043074 fnInt(y1,x,-7,0) 2.46944444447 evalF nDer der1 der2 fnInt▶

- If you evaluate a definite integral using antiderivative formulas and check your answer with the calculator using fnInt, you may sometimes find a slight difference in the last few decimal places. Remember, the TI-85 and the TI-86 are evaluating the definite integral using an approximation technique.

6.6.2 EVALUATING A DEFINITE INTEGRAL FROM THE GRAPHICS SCREEN

Provided that a and b are possible x-values when you trace the graph, you can find the value of the definite integral $\int_{a}^{b} f(x)\, dx$ from the graphics screen. (For "nice" trace numbers, you can often find the exact x-value you need if you graph in the ZDECM screen or set the viewing window so that xMax − xMin equals a multiple of 12.6.)

Have the model for the average sea level in y1 (see Section 6.6.1). Suppose we want to find $\int_{-7}^{0} r(t)\, dt$.	

TI-85 Because ⁻7 is not a possible trace value in the window set by the stat plot, change **xMin** to ⁻7 and **xMax** to 0.

Graph **y1**. Press [MORE] [F1] (MATH) [F5] (∫f(x)dx). The blinking cursor means the TI-85 is waiting for you to give it the lower limit on the integral. Press and hold ◀ until you reach ⁻7.

Press [ENTER] to set the lower limit. The calculator is now waiting for you to give it the upper limit on the integral.

Press and hold ▶ until you reach X = 0.

Press [ENTER] to calculate the numerical value of the definite integral.

TI-86 This calculator allows you to input any value of *x* between **xMin** and **xMax** when finding the value of a definite integral from the graphics screen.

Have all **STAT PLOTS** turned off. Set the window shown on the right -- the scatter plot window for the average sea level data with **xMax** reset to 0.

Draw the graph of *r(t)* . Press [MORE] [F1] (MATH) [F3] (∫f(x)dx). When the TI-86 asks for a **lower limit**, simply type ⁻7 and then press [ENTER].

When the calculator asks for an **upper limit**, type 0 and then press [ENTER].

Note that if you had not previously changed **xMax** from ⁻0.4 to 0, you would have gotten an ERR: INVALID message at this point because 0 would not have been a value included between **xMin** and **xMax**.

The value of the integral $\int_{-7}^{0} r(t)\, dt$ is displayed and the two regions are shaded.

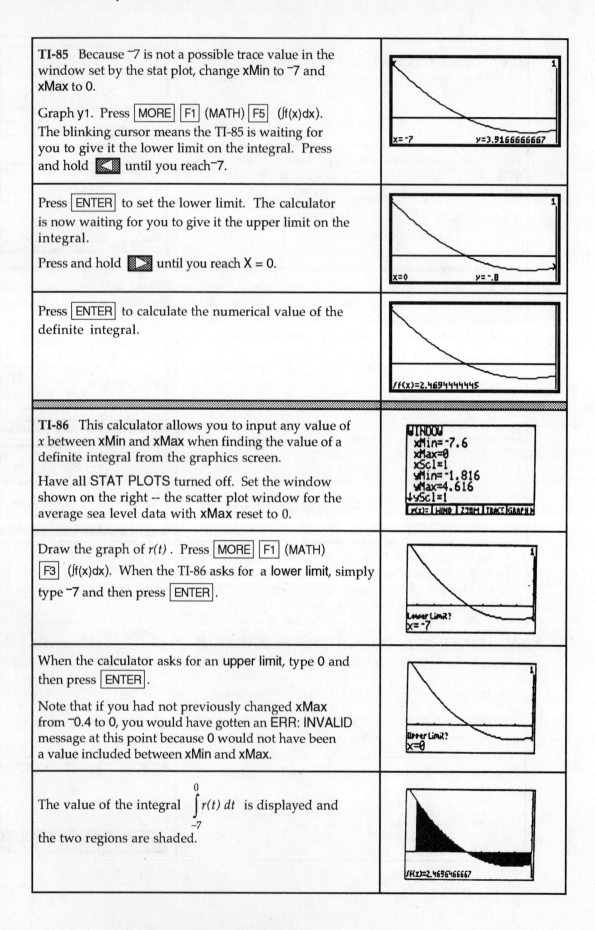

Chapter 7 Analyzing Accumulated Change: More Applications of Integrals

📖 7.1 Differences of Accumulated Changes

This chapter helps you effectively use your calculator's numerical integrator with various applications pertaining to the accumulation of change. In this first section, we focus on the differences of accumulated changes.

7.1.1 FINDING THE AREA BETWEEN TWO CURVES

Finding the area of the region enclosed by two functions uses many of the techniques presented in preceding sections. Suppose we want to find the difference between the accumulated change of $f(x)$ from a to b and the accumulated change of $g(x)$ from a to b where $f(x) = 0.3x^3 - 3.3x^2 + 9.6x + 3.3$ and $g(x) = -0.15x^2 + 2.03x + 3.33$. The input of the leftmost point of intersection of the two curves is a and the input of the rightmost point of intersection is b.

Enter $f(x)$ in y1 and $g(x)$ in y2 in the y(x)= list. From Figure 7.9 in the text, we see that xMin = 1, xMax = 7, yMin = 0, and yMax = 12. Set these (RANGE or WIND) values and then graph the two functions. Next, find the two intersection points. (The x-values of these points of intersection will be the limits on the integrals we use to find the area.)	![graph with y(x)= RANGE ZOOM TRACE GRAPH]
TI-85 Press MORE F1 (MATH) MORE F5 (ISECT) **TI-86** Press MORE F1 (MATH) MORE F3 (ISECT) **Both** Use ◀ and/or ▶ to move near the point of intersection in the middle of the screen. Press ENTER each time the calculator pauses to tell it the first curve, the second curve, and your guess for the point of intersection.	![graph ISECT X=3.715045B525 y=B.8013082277]
The leftmost visible point of intersection is displayed. To avoid making a mistake copying the x-value and to eliminate as much rounding error as possible, return to the home screen and store this value in A with the keystrokes x-VAR STO▸ A	x→A 3.71504585253
Press GRAPH, and repeat the process to find the right-most intersection point. Do not forget to move the cursor near the intersection point on the right after pressing the ISECT key. Store the x-value in B.	![graph ISECT X=6.7B09845811 y=10.198135916]

The combined area of the three regions is

$$\int_1^A (f-g)\,dx \; + \; \int_A^B (g-f)\,dx + \int_B^7 (f-g)\,dx.$$

(If you are not sure which curve is on top in each of the various regions, trace the curves.)

Use either the home screen or the graphics screen to find the value of the definite integral

$$\int_1^7 (f-g)\,dx \; - \; \int_1^7 \big(f(x)-g(x)\big)\,dx.$$

Note that the value of the integral is *not* the same as the area between the two curves.

7.3 Streams in Business and Biology

You will find your calculator very helpful when dealing with streams that are accumulated over finite intervals. However, because your calculator's numerical integrator evaluates only definite integrals, you cannot use it to find the value of an improper integral.

7.3.1 FUTURE VALUE OF A DISCRETE INCOME STREAM

We use the sequence command to find the future value of a discrete income stream. The change in the future value at the end of T years that occurs because of a deposit of $\$A$ at time t where interest is earned at an annual rate of $100r\%$ compounded n times a year is

$$f(t) = A\left(1+\frac{r}{n}\right)^{n(T-t)} \quad \text{dollars per compounding period}$$

where t is the number of years since the first deposit was made. We assume the initial deposit is made at time $t = 0$ and the last deposit is made at time $T - \frac{1}{n}$. The increment for a discrete stream involving n compounding periods and deposits of $\$A$ at the beginning of each compounding period is $\frac{1}{n}$. Thus, the sequence command for finding the future value of the discrete income stream is *seq(f(x), x, 0, T−1/n, 1/n)*.

Suppose that you invest $75 each month in a savings account yielding 6.2% APR compounded monthly. What is the value of your savings in 3 years? To answer this question, note that the change in the future value that occurs due to the deposit at time t is

$$f(t) = 75\left(1+\frac{0.062}{12}\right)^{12(3-t)} \approx 75\big(1.005166667^{12}\big)^{(3-t)} \approx 75(1.06379)^{(3-t)}$$

Now, find the future value of this stream with $A = 75$, $r = 0.062$, $T = 3$, and $n = 12$.

Enter $f(t)$, using x as the input variable, in location y1 of the y(x)= list. Note that if you want an exact answer you should enter the following:

    y1 = 75(1+ 0.062/12)^(12(3−x))

(You must carefully use parentheses with this form of the equation.)

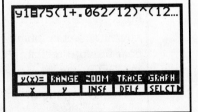

Return to the home screen. Enter seq (y1, x, 0, 3 –1/12, 1/12) . (Note that since you start counting at 0, the ending value will be one increment less than the number of years the money accumulates in the account.)	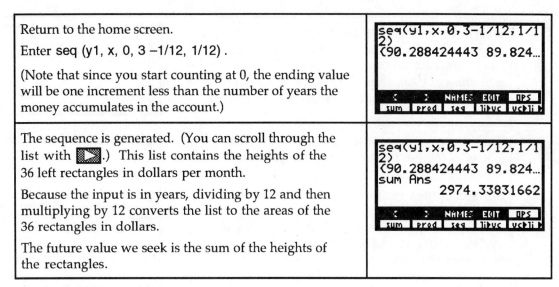
The sequence is generated. (You can scroll through the list with 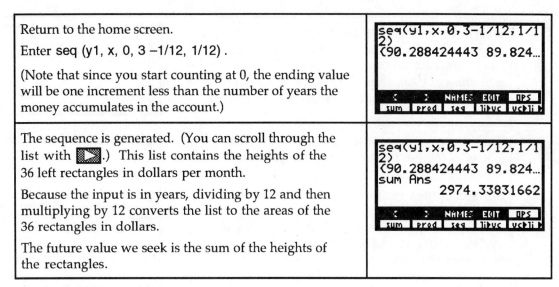.) This list contains the heights of the 36 left rectangles in dollars per month. Because the input is in years, dividing by 12 and then multiplying by 12 converts the list to the areas of the 36 rectangles in dollars. The future value we seek is the sum of the heights of the rectangles.	

- When using definite integrals to approximate either the future value or the present value of a discrete income stream or to find the future value or the present value of a continuous stream, use the fnInt command to find the area between the appropriate continuous rate of change function and the *t*-axis from 0 to T.

7.4 Integrals in Economics

Consumers' and producers' surplus, being defined by definite integrals, are easy to find using the calculator. You should always draw graphs of the demand and supply functions and think of the surpluses in terms of area to better understand the questions being asked.

7.4.1 CONSUMERS' SURPLUS   Suppose that the demand for mini-vans in the United States can be modeled by $D(p) = 14.12(0.933)^p - 0.25$ million mini-vans when the market price is p thousand dollars per mini-van.

At what price will consumers purchase 2.5 million mini-vans? We solve $D(p) = 2.5$ to find the price. Enter $14.12(0.933)^p - 0.25$ in y1. (You could use the solver on the home screen, but we intend to graph $D(p)$. Don't forget to use x as the input variable.)	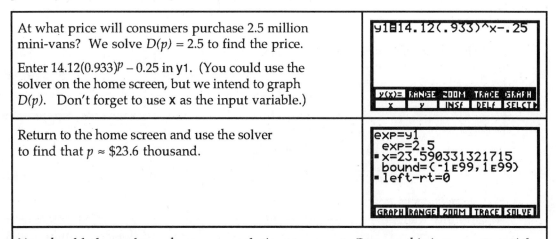
Return to the home screen and use the solver to find that $p \approx \$23.6$ thousand.	
You should always know how many solutions to expect. Because this is an exponential equation, it can have no more than one solution. Refer to Section 1.2.2 of this *Guide* for instructions on using the solver.	

Now, let's find if the model indicates a possible price above which consumers will purchase no mini-vans. If so, we will find that price.

You should have the demand function in y1. Graph y1 in a suitable viewing window. Since the output is mini-vans, we set xMin = 0 and consider the graph for different values of xMax. First, set xMax = 15 (or some other value) and use ZFIT to draw the graph.

Because we are trying to see if the demand function crosses or touches the input axis, reset yMin to ⁻5 and xMax to a larger value, say 65. Use F3 (GRAPH) to graph the demand function.

It is difficult to tell if there is an intercept. So, use MATH (ROOT) to find, should it exist, the root of the demand function.

Instead of drawing the graph, you could have used the solver to find the location of the x-intercept. Notice that here we are looking for where y1 = 0.

Consumers will not pay more than about $58,200 per mini-van.

What is the consumers' surplus when 2.5 million mini-vans are purchased? We earlier found that the market price for this quantity is 23.59033 thousand dollars.

 TI-85 The consumers' surplus is *approximately* the area of the region shown to the right (the trace value on the calculator can not set exactly to the market price).

TI-86 You can type in the values for the upper and lower limits and obtain a much better approximation.

(The TI-85 does not shade the region when it finds the value of the integral.)

Find the consumers' surplus, which is the area of the shaded region, by finding the value of

$$\int_{M}^{P} D(p)\, dp = \int_{23.59033132}^{58.16700863} \left(14.12(0.933)^p\right) dp.$$

(Remember that you can avoid retyping the long decimal numbers by storing them to various memory locations as you find them.)

- Carefully watch the units involved in your computations. Refer to the statement of the problem and note that the height is measured in thousand dollars per mini-van and the width in million mini-vans. Thus, the area should be written in units that make sense in the context of this problem:

$$\text{height} * \text{width} \approx 27.4 \left(\frac{\text{thousand dollars}}{\text{minivan}}\right)(\text{million minivans}) = \$\,27.4 \text{ thousand million}$$

$$= \$27.4\,(1{,}000)(1{,}000{,}000) = \$27.4(10^9) = \$27.4 \text{ billion dollars}$$

7.4.2 PRODUCERS' SURPLUS   When dealing with supply functions, use definite integrals in a manner similar to that for consumers' quantities to find producers' revenue, producers' surplus, and so forth.   To illustrate, suppose the demand for mini-vans in the United States can be modeled by $D(p) = 14.12(0.933)^p - 0.25$ million mini-vans and the supply curve is $S(p) = 0$ million mini-vans when $p < 15$ and $S(p) = 0.25p - 3.75$ million mini-vans when $p \geq 15$ where the market price is p thousand dollars per mini-van.

Enter the demand curve in **y1** and the supply curve in **y2**. (Remember to use **x** as the input variable.) Draw a graph of the demand and supply functions in an appropriate window, say $0 \leq x \leq 65$ and $^-5 \leq y \leq 16$.	![y1=14.12(.933)^x-.25 y2=0(x<15)+(.25 x-3....]
Market equilibrium occurs when $D(p) = S(p)$. Use the methods of Section 7.1.1 of this *Guide* to find the intersection of the two curves. Store the *x*-value of the intersection as *B*.	![ISECT X=24.399630711 Y=2.3499076777]
Producers' surplus is found by evaluating $\displaystyle\int_{15}^{B} S(p)\,dp$	![x→B 24.3996307107 fnInt(y2,x,15,B) 11.0441321873]

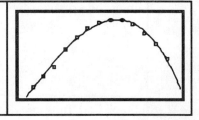

📖 7.5 Average Value and Average Rates of Change

You need to carefully read any question involving average value in order to determine which quantity is involved. Considering the units of measure in the situation can be of tremendous help when trying to determine which function to integrate when finding average value.

7.5.1 AVERAGE VALUE OF A FUNCTION   Suppose that the hourly temperatures shown below were recorded from 7 a.m. to 7 p.m. one day in September.

Time	7am	8	9	10	11	noon	1pm	2	3	4	5	6	7
Temp. (°F)	49	54	58	66	72	76	79	80	80	78	74	69	62

Enter the input data in **L1** as the number of hours after midnight: 7 am is 7 and 1 pm is 13, etc.

First, we fit a cubic model $\quad t(h) = ^-0.03526h^3 + 0.71816h^2 + 1.584h + 13.689$ °F where h is the number of hours after midnight. Graph this model on the scatter plot of the data and see that it provides a good fit.	

Next, you are asked to calculate the average temperature. Because temperature is measured in this example in degrees Fahrenheit, the units on your result should be °F. When evaluating integrals, it helps to think of the units of the integration result as (height)(width) where the height units are the output units and the width units are the input units of the function that you are integrating. That is,

$$\int_{9\text{ hours}}^{18\text{ hours}} t(h) \text{ degrees } dh \quad \text{has units of (degrees)(hours)}.$$

When we find the average value, we divide the integral by (upper limit − lower limit). So,

$$\text{average value} = \frac{\int_{9\text{ hours}}^{18\text{ hours}} t(h) \text{ degrees } dh}{18\text{ hours} - 9\text{ hours}} = \frac{(T(18) - T(9)) \text{ degrees} \cdot \text{hours}}{9\text{ hours}}$$

where $T(h)$ is an antiderivative of $t(h)$. Because the "hours" cancel, the result is in degrees as is desired. Remember, when finding average value, *the units of the average value are always the same as the output units of the quantity you are integrating.*

Find the average value of the temperature between 9 a.m. and 6 p.m. to be approximately 74.4 °F.	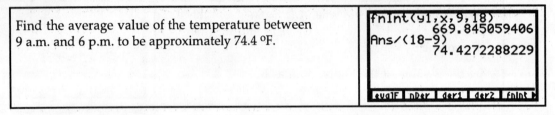

The third part of this example asks you to find the average rate of change of temperature from 9 a.m. to 6 p.m. Again, let the units be your guide. Because temperature is measured in °F and input is measured in hours, the average rate of change is measured in output units per input units = °F per hour. Thus, we find the average rate of change to be

$$\text{average rate of change of temperature} = \frac{(t(18) - t(9)) \text{ °F}}{(18 - 9) \text{ hours}} = 0.98 \text{ °F per hour}$$

7.5.2 GEOMETRIC INTERPRETATION OF AVERAGE VALUE

What does the average value of a function mean in terms of the graph of the function? Consider the model we found above for the temperature one day in September between 7 a.m. and 7 p.m.:

$$t(h) = -0.03526h^3 + 0.71816h^2 + 1.584h + 13.689 \text{ degrees } t \text{ hours after 7 a.m.}$$

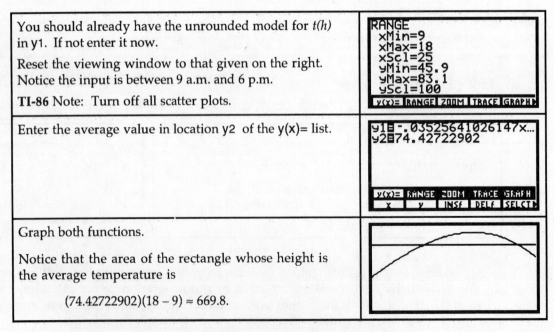

You should already have the unrounded model for $t(h)$ in y1. If not enter it now.

Reset the viewing window to that given on the right. Notice the input is between 9 a.m. and 6 p.m.

TI-86 Note: Turn off all scatter plots.

Enter the average value in location y2 of the y(x)= list.

Graph both functions.

Notice that the area of the rectangle whose height is the average temperature is

$$(74.42722902)(18 - 9) \approx 669.8.$$

The area of the rectangle equals the area of the region below the temperature function *t(h)* and above the *t*-axis between 9 a.m. and 6 p.m.

(In this application, the area does not have a meaningful interpretation because its units are (degrees)(hours).)

📖 7.6 Probability Distributions and Density Functions

Most of the applications of probability distributions and density functions use technology techniques that have already been discussed. Probabilities are areas whose values can be found by integrating the appropriate density function. A cumulative density function is an accumulation function of a probability density function.

You will find your calculator's numerical integrator especially useful when finding means and standard deviations for some probability distributions because those integrals often contain expressions for which we have not developed an algebraic technique for finding an antiderivative.

7.6.1 NORMAL PROBABILITIES

The normal density function is the most well known and widely used probability distribution. If you are told that a random variable *x* has a normal distribution *N(x)* with mean μ and standard deviation σ, the probability that *x* is between *a* and *b* is

$$\int_a^b N(x)\,dx = \int_a^b \frac{1}{\sigma\sqrt{2\pi}} e^{\frac{-(x-\mu)^2}{2\sigma^2}}\,dx.$$

Suppose a light-bulb manufacturer advertises that the average life of the company's soft-light bulb is 900 hours with a standard deviation of 100 hours. Suppose also that we know the distribution of the life of these bulbs, with the life span measured in hundreds of hours, is modeled by a normal density function. Find the probability that any one of these light bulbs last between 900 and 1000 hours.

We are told that $\mu = 9$ and $\sigma = 1$. (Note that *x* is measured in hundreds of hours, so we must convert to these units.) Enter $$N(x) = (1/\sqrt{(2\pi)})e^{(-(x-9)^2/2)}$$ in y1 using $\mu = 9$ and $\sigma = 1$. (Very carefully watch your use of parentheses.)	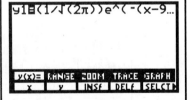
Find $P(9 < x < 10)$. Graph *N(x)*, remembering that nearly all the area is between $\mu-3\sigma$ and $\mu+3\sigma = $ [6, 12]. Use **ZFIT** to graph.	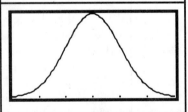
With the graph on the screen, use MORE F1 (MATH) F3 ($(\int f(x)dx)$, choose 9 as the lower limit, choose 10 as the upper limit, and see the (shaded on the TI-86) region whose area is $P(9 < x < 10)$.	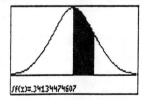

Chapter 8 Repetitive Change: Cycles and Trigonometry

📖 8.1 Functions on Angles: Sines and Cosines

Before you begin this chapter, go back to the first page of this *Guide* and check the basic setup, the statistical setup, and the window setup. If these are not set as specified in Figures 1, 2, and 3, you will have trouble using your calculator in this chapter. Pay careful attention to the third line in the MODE screen in the basic setup. The Radian/Degree mode setting affects the calculator's interpretation of the ANGLE menu choices. Your calculator's MODE menu should always be set to Radian unless otherwise specified.

8.1.1 CONVERTING ANGLES FROM RADIANS TO DEGREES   Even though the process of converting angles from one measure to another is a simple arithmetic task, we cover it here to show you more of your calculator's functionality. Suppose you want to convert $7\pi/6$ to degree measure. (Remember that when no units are specified, the angle is assumed to be expressed in radian measure.)

As shown in the text example, convert $7\pi/6$ radians to degrees with multiplication by the proper factor. (Let the units be your guide and carefully use parentheses.) $$\left(\frac{7\pi}{6} \text{ radians}\right)\left(\frac{180^\circ}{\pi \text{ radians}}\right) = 210^\circ$$	`(7π/6)(180/π)` `                210`

If you want to use your calculator for this conversion *to* degrees, first set the calculator to degree mode. Press 2nd MORE (MODE), choose Degree in the third line and press ENTER. Return to the home screen.	`Normal Sci Eng`
	`Float 012345678901`
	`Radian Degree`
	`RectC PolarC`
	`Func Pol Param DifEq`
	`Dec Bin Oct Hex`
	`RectV CylV SphereV`
	`dxDer1 dxNDer`

Tell the calculator the angle and that it is measured in radians with (7 2nd ^ (π) ÷ 6) 2nd X (MATH) F3 (ANGLE) F2 (ʳ). Press ENTER to convert the angle to degrees.	`(7π/6)ʳ`
	`                210`
	
	`NUM  PROB ANGLE HYP  MISC`
	`  °    r    '  ▸DMS`

While it is not necessary in this example, if you have an angle that you want expressed in degrees, minutes, and seconds, first convert the angle to degrees and then press 2nd X (MATH) F3 (ANGLE) F4 (▸DMS). (Don't forget to set your calculator mode back to radian measure when you finish the conversions!)	`(7π/6)ʳ`
	`                210`
	`Ans▸DMS`
	`           210°0'0"`
	
	`NUM  PROB ANGLE HYP  MISC`
	`  °    r    '  ▸DMS`

8.1.2 CONVERTING ANGLES FROM DEGREES TO RADIANS   Converting angles from degree measure to radian measure is similar to process described above. However, since you wish to convert to radian measure, the mode setting should be Radian. Suppose that you need to convert 135° to radian measure.

As shown in the text example, do this numerically by multiplication of the proper factor. (Let the units be your guide and carefully use parentheses.) $$\left(135°\right)\left(\frac{\pi \text{ radians}}{180°}\right) = \frac{3\pi}{4} \text{ radians} = \frac{3\pi}{4}$$	`135(π/180)`
	`          2.35619449019`
	`3π/4`
	`          2.35619449019`

Because you want to convert *to* radians, press MODE, choose Radian in the third line and press ENTER.
 	
Enter the degree measure of the angle and tell the calculator the angle is in degrees with 2nd X (MATH) F3 (ANGLE) F1 (°). ENTER converts to radians.
	`135°`
	`          2.35619449019`
	
	`NUM  PROB ANGLE HYP  MISC`
	`  °    r    '  ▶DMS`

8.1.3 EVALUATING TRIGONOMETRIC FUNCTIONS   Just as it was important to have the correct mode set when changing from degrees to radians or vice versa, it is essential that you have the correct mode set when evaluating trigonometric function values. Unless you see the degree symbol, the mode setting should be Radian.

Find $\sin \frac{9\pi}{8}$ and $\cos \frac{9\pi}{8}$.
 	
Because these angles are in radians, be certain that Radian is chosen in the third line of the Mode screen. The SIN and COS keys are above the , and (keys on your calculator's keyboard.
	`sin (9π/8)`
	`         -.382683432365`
	`cos (9π/8)`
	`         -.923879532511`

It is also essential that you use parentheses -- the calculator's order of operations are such that the sine function takes precedence over division.
 	
Without telling your calculator to first divide, it evaluates $\frac{\sin (9\pi)}{8}$ and $\frac{\cos (9\pi)}{8}$.
	`         -.382683432365`
	`cos (9π/8)`
	`         -.923879532511`
	`sin 9π/8`
	`              2.5E-14`
	`cos 9π/8`
	`                -.125`

The values of the trigonometric functions are the same regardless of whether the angles are expressed in radian measure or degree measure. To verify this, first convert the angle in radians to degree measure, and then re-evaluate the sine and cosine of the angle.

Change the mode setting to Degree and then convert $\frac{9\pi}{8}$ to degrees. Evaluate the sine and cosine functions.
 	
(When you finish, don't forget to change the mode setting back to Radian!)
	`(9π/8)ʳ`
	`                202.5`
	`sin 202.5`
	`         -.382683432365`
	`cos 202.5`
	`         -.923879532511`

8.2 Cyclic Functions as Models

We now introduce another model -- the sine model. As you might expect, this model is fit to data that repeatedly varies between alternate extremes. The form of the model is $f(x) = a \sin(bx + c) + d$ where a is the amplitude, $2\pi/|b|$ is the period, $\bar{}c/b$ is the horizontal shift, and d is the vertical shift.

The sine model is available in the TI-86, but on the TI-85, we use program[1] SINREG to fit the sine model. (SINREG is given in the TI-85/86 Appendix.)

8.2.1 FITTING A SINE MODEL TO DATA

Before fitting any model to data, remember that you should construct a scatter plot of the data and observe what pattern the data appears to follow. We illustrate finding a sine model for cyclic data with the hours of daylight on the Arctic Circle as a function of the day of the year on which the hours of daylight are measured. (January 1 is day 1.)

Day of the year	$\bar{}10$	81.5	173	264	355	446.5	538	629	720	811.5
Hours of daylight	0	12	24	12	0	12	24	12	0	12

Enter these data with the day of the year in L1 and the hours of daylight in L2.

Construct a scatter plot of the data. It appears to be cyclic. Either look at the data, move around the screen with the arrow keys on the TI-85, or use TRACE on the TI-86 to measure the horizontal distance between one high point and the next (or between two successive low points, etc.). One cycle of the data appears to be about 365 days.	x=541.57857143 y=23.980645161 period ≈ 542– 177
TI-85 Run program SINREG. The program's initial message reminds you that the input data should be in list L1 and the output data in list L2. *The input data should be in order, from smallest to largest, when using this program.* You must supply the program with the period of the model and the maximum iterations -- the number of times the program should cycle thorough the process of determining the best-fit model.	Period Guess? 365 Maximum Iteration? 3 The more iterations you use, the longer it takes the program to run. Try a value between 3 and 10 for best results.
If the program finds a good fit before it completes the maximum number of iterations, it stops. For instance, these data are so close to a sine curve that it took only 1 iteration. (You need to look at the program to know this, however.) The program stores the model in y1, so all you need do is run program STPLT to graph the model on the scatter plot of the data.	MODEL Y=Asin (BX+C)+D A=12.0001111242 B=.017214206321 C=⁻1.40295781516 D=12 (If the graph of the model looks like a line, you have forgotten to put your calculator back in radian mode!)

---

[1]This program is based on a program that was modified by John Kenelly from material by Charles Scarborough of Texas Instruments. The authors sincerely thank Robert Simms for this help with this version of the program.

TI-86 The sine model SinR on the TI-86 is in the list with the other models in the STAT (CALC) menu.
 	
Enter the list containing the input data, the list containing the output data, and the location of the y(x)= list that you wish the model to appear. Press ENTER .
	```SinR L1,L2,y1```
	
	`CALC EDIT PLOT DRAW VARS`
	`PwrR SinR LgstR P2Reg P3Reg`

The sine model of best fit appears on the screen and is pasted into the y1 location of the y(x)= list.
 	
(If you wish, you can tell the calculator how many times to go through the routine that fits the model. This number of iterations is 3 if not specified. The number should be typed before L1 when initially finding the model.)
	```SinReg```
	`y=a*sin(bx+c)+d`
	`PRegC=`
	`{12.0001111242 .0172…`
	
	`CALC EDIT PLOT DRAW VARS`
	`PwrR SinR LgstR P2Reg P3Reg`

Press GRAPH F5 (GRAPH) to graph the model on the scatter plot of the data. (If the graph of the model looks like a line, you have forgotten to put your calculator back in radian mode!)	
Even though it did not occur in this example, you may get a **SINGULAR MATRIX** error when trying to fit a sine model to certain data. If so, try specifying an estimate for the period of the model. Recall that our estimate of the period is 365 days.	```SinR L1,L2,365,y1```

Notice how much faster the TI-86 finds the model!
 	
(If you do not think the original model the calculator finds fits the data very well, try specifying a period and see if a better-fitting model results. It didn't here, but it might with a different set of data.)
	```SinReg```
	`y=a*sin(bx+c)+d`
	`PRegC=`
	`{12.0001111242 .0172…`

8.3 Rates of Change and Derivatives

All the previous techniques given for other models also hold for the sine model. You can find intersections, maxima, minima, inflection points, derivatives, integrals, and so forth.

8.3.1 DERIVATIVES OF SINE AND COSINE MODELS

Evaluate der1 at a particular input to find the value of the derivative of the sine model at that input. Suppose the calls for service made to a county sheriff's department in a certain rural/suburban county can be modeled as $c(h) = 2.8 \sin(0.262h + 2.5) + 5.38$ calls during the hth hour after midnight.

Enter the model in some location of the y(x)= list, say y1. In another location, say y2, enter the calculator's numerical derivative.	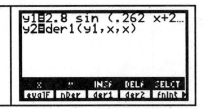

Even though your calculator will not give you an equation for the derivative, you can use one of the methods illustrated in Sections 4.4.1 and/or 4.4.2 of this *Guide* to check the derivative you obtain using derivative rules.

Enter your derivative formula in y3, turn off y1, and use the table to compare several values of the calculator's derivative (y2) and your derivative (y3).

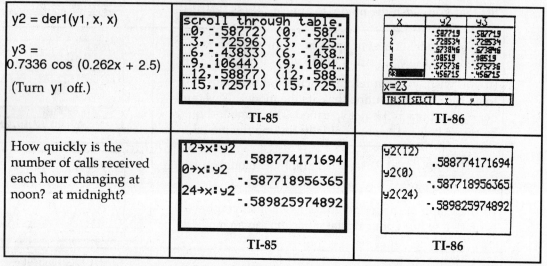

| y2 = der1(y1, x, x)

y3 =
0.7336 cos (0.262x + 2.5)

(Turn y1 off.) | scroll through table.
...0, -.58772) (0, -.587...
...3, -.72596) (3, -.725...
...6, -.43833) (6, -.438...
...9, .10644) (9, .1064...
...12, .58877) (12, .588...
...15, .72571) (15, .725...
TI-85 | TI-86 table with x, y2, y3 columns
x=23
TI-86 |
| How quickly is the number of calls received each hour changing at noon? at midnight? | 12→x: y2
　　　　.588774171694
0→x: y2
　　　-.587718956365
24→x: y2
　　　-.589825974892
TI-85 | y2(12)
　　　　.588774171694
y2(0)
　　　-.587718956365
y2(24)
　　　-.589825974892
TI-86 |

- The calculator screens shown above show that to answer these questions, simply evaluate y2 (or your derivative in y3) at 12 for noon and 0 (or 24) for midnight. Also, you weren't told if "midnight" refers to the initial time or 24 hours after that initial time.

📖 8.5 Accumulation in Cycles

As with the other models we have studied, applications of accumulated change with the sine and cosine models involve the calculator's numerical integrator fnInt.

8.5.1 INTEGRALS OF SINE AND COSINE MODELS   Suppose that the rate of change of the temperature in Philadelphia on August 27, 1993 can be modeled as

$$t(h) = 2.733 \cos(0.285h - 2.93) \text{ °F per hour}$$

h hours after midnight. Find the accumulated change in the temperature between 9 a.m. and 3 p.m. on August 27, 1993.

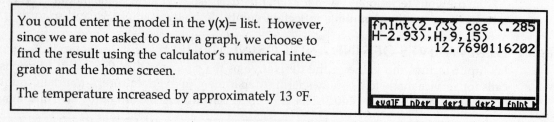

| You could enter the model in the y(x)= list. However, since we are not asked to draw a graph, we choose to find the result using the calculator's numerical integrator and the home screen.

The temperature increased by approximately 13 °F. | fnInt(2.733 cos (.285
H-2.93),H,9,15)
　　　　12.7690116202

evalF　nDer　der1　der2　fnInt ▶ |

Chapter 9   Ingredients of Multivariable Change: Models, Graphs, Rates

📖 9.1 Cross-Sectional Models and Multivariable Functions

For a multivariable function with two input variables described by data given in a table with the values of the *first* input variable listed *horizontally* across the top of the table and the values of the *second* input variable listed *vertically* down the left side of the table, obtain cross-sectional models by entering the appropriate data in lists L1 and L2 and then proceeding as indicated in Chapter 1 of this *Guide*.

9.1.1 FINDING CROSS-SECTIONAL MODELS (holding the first input variable constant)
Using the elevation data in Table 9.1 of the text, find the cross-sectional model *E(0.8, n)* as described below. Remember that "rows" go from left to right horizontally and "columns" go from top to bottom vertically.

When considering *E(0.8, n)*, *e* is constant at 0.8 and *n* varies.

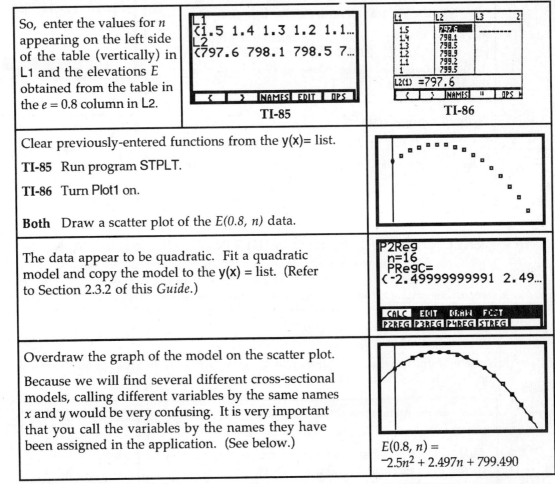

So, enter the values for *n* appearing on the left side of the table (vertically) in L1 and the elevations *E* obtained from the table in the *e* = 0.8 column in L2.

Clear previously-entered functions from the y(x)= list.

TI-85 Run program STPLT.

TI-86 Turn Plot1 on.

Both Draw a scatter plot of the *E(0.8, n)* data.

The data appear to be quadratic. Fit a quadratic model and copy the model to the y(x) = list. (Refer to Section 2.3.2 of this *Guide*.)

Overdraw the graph of the model on the scatter plot.

Because we will find several different cross-sectional models, calling different variables by the same names *x* and *y* would be very confusing. It is very important that you call the variables by the names they have been assigned in the application. (See below.)

$E(0.8, n) =$
$-2.5n^2 + 2.497n + 799.490$

- Remember that when finding or graphing a model, your calculator always calls the input variable x and the output variable y. When working with multivariable functions, you must translate the calculator's model y1 ≈ ⁻2.5x^2 + 2.497x + 799.490 to the symbols used in this problem to obtain $E(0.8, n) = -2.5n^2 + 2.497n + 799.490$.

9.1.2 FINDING CROSS-SECTIONAL MODELS (holding the second input variable constant)

Using the elevation data in Table 9.1 of the text, find the cross-sectional model $E(e, 0.6)$ as discussed below. When considering $E(e, 0.6)$, n is constant at 0.6 and e varies.

Thus, enter the values for e appearing across the top of the table (horizontally) in L1 and the elevations E obtained from the table in the $n = 0.6$ row in L2.	L1 {0 .1 .2 .3 .4 .5 .6... L2 {802.8 801.6 800.8 8... 〈 〉 NAMES EDIT OPS **TI-85**
	L1 L2 L3 2 0 802.8 .1 801.6 .2 800.8 .3 800.3 .4 800 .5 799.9 L2(1) =802.8 〈 〉 NAMES " OPS 〉 **TI-86**
You can either turn y1 "off" by placing the blinking cursor over the equals sign and pressing ENTER or you can delete it. Next, draw a scatter plot of the $E(e, 0.6)$ data.	
The data appears to be cubic. Fit a cubic model and copy it to the y2 location of the y(x)= list. Be certain that you translate the calculator's model $y2 \approx {}^{-}10.124X^{\wedge}3 + 21.347X^{\wedge}2 - 13.972X + 802.809$ to the symbols used in the example. (See the note at the end of Section 9.1.1.)	y1={}^-2.49999999991x^2... y2▪{}^-10.124103599x^3+... y(x)= RANGE ZOOM TRACE GRAPH x y INSf DELf SELCT〉
Write this cross-sectional model as $E(e, 0.6) = {}^{-}10.124e^3 + 21.347e^2 - 13.972e + 802.809$ Overdraw the graph of the model on the scatter plot.	

9.1.3 EVALUATING OUTPUTS OF MULTIVARIABLE FUNCTIONS

Outputs of multivariable functions are found by evaluating the function at the given values of the input variables. One way to find multivariable function outputs is to evaluate them on the home screen. For instance, the compound interest formula tells us that

$$A(P, t) = P\left(1 + \frac{0.06}{4}\right)^{4t} \text{ dollars}$$

is the amount to which P dollars will grow over t years in an account paying 6% interest compounded quarterly. Let's review how to find $A(5300, 10)$.

Type, on the home screen, the formula for $A(P, t)$, substituting in $P = 5300$ and $t = 10$. Press ENTER. Again, be warned that you must carefully use the correct placement of parentheses. Check your result with the values in Table 9.4 in the text to see if $9614.30 is a reasonable amount.	5300(1+.06/4)^(4*10) 9614.29756594

Even though it is not necessary in this example, you may encounter activities in this section in which you need to evaluate a multivariable function at several different inputs. Instead of individually typing each one on the home screen, there are easier methods. You will also use the techniques shown below in later sections of this chapter.

When evaluating a multivariable function at several different input values, you may find it more convenient to enter the multivariable function in the graphing list. It is very important to note at this point that while we have previously used x as the input variable when entering functions in the y(x)= list, we do *not* follow this rule when *evaluating* functions with more than one input variable.

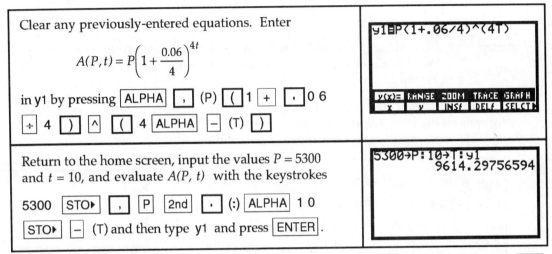

Clear any previously-entered equations. Enter

$$A(P,t) = P\left(1 + \frac{0.06}{4}\right)^{4t}$$

in y1 by pressing ALPHA , (P) (1 + . 0 6 ÷ 4) ^ (4 ALPHA − (T))

Return to the home screen, input the values $P = 5300$ and $t = 10$, and evaluate $A(P, t)$ with the keystrokes

5300 STO▸ , P 2nd . (:) ALPHA 1 0 STO▸ − (T) and then type y1 and press ENTER .

- **TI-86 Note**: Because the TI-86 can recall many previously-entered expressions with 2nd ENTER (ENTRY), this storing process can be done in several different steps.

To find another output for this function, say $A(8700, 8.25)$, repeat the above procedure:

Bring back the last line by pressing 2nd ENTER (ENTRY). Edit the expression to input the values $P = 8700$ and $t = 8.25$. Press ENTER to evaluate.

Warning: Recall that the your calculator recognizes only x as an input variable in the y(x)= list when you *graph* and when you use the *table*. Thus, you should not attempt to graph the current y1 = A(P, t) or use the table because the calculator considers this y1 a constant. (Check it out and see that the graph is a straight line at about 14220 and that all values in the table are approximately 14220.)

📖 9.2 Contour Graphs

Because any program that we might use to graph[1] a three-dimensional function would be very involved to use and take a long time to execute, we do not attempt to do such. Instead, we discover graphical information about three-dimensional graphs using their associated contour curves.

---

[1]A program to graph a multivariable function with two input variables (as well as many other Texas Instruments programs, are available at http://www.ti.com/calc/docs/80xthing.htm.

9.2.1 SKETCHING CONTOUR CURVES

When you are given a multivariable function model, you can draw contour graphs for functions with two input variables using the three-step process described below. We illustrate by drawing the $520 constant-contour curve with the function for the monthly payment on a loan of A thousand dollars for a period of t years:

$$M(t, A) = \frac{5.833333A}{1 - 0.932583^t} \text{ dollars}$$

Step 1: Set $M(t, A) = 520$. Since we intend to use the solve routine to find the value of A at various values of t, write the function as

$$M(t, A) - 520 = 0.$$

Step 2: Choose values for t and solve for A to obtain points on the $520 constant-contour curve. Obtain guesses for the values of t from the table of loan amounts in the table in Figure 9.14 of the text.

You can either enter the $M(t, A)$ formula in y1 or enter it on the home screen. If you enter the formula in the graphing list, be certain to use the letters **A** and **T** and to indicate for which variable you are solving when using the solve routine.

Access the **SOLVER** with 2nd ENTER (SOLVER), and clear any previously-entered equations with CLEAR. Enter $M(t, A)$ in the **eqn** location with the keystrokes 5.833333 ALPHA A ÷ ((1 − .932583 ^ ALPHA T) .	`eqn:…3A/(1-.932583^T)` `eqn  y1` (If necessary, refer to Section 1.2.2 of this *Guide* to review the instructions for using the solver.)
Press ENTER. Since $M(t, A) = 520$, enter 520 in the **eqn** value location. Choose a value for t, say $t = 12$ years and solve for A.	`exp=5.833333A/(1-.93…` `exp=520` `▪A=50.565146552928` `T=12` `bound=(-1E99,1E99)` `▪left-rt=0` `GRAPH RANGE ZOOM TRACE SOLVE`
Choose another value for t, say $t = 16$ years. Solve for A.	`exp=5.833333A/(1-.93…` `exp=520` `▪A=59.962781721818` `T=16` `bound=(-1E99,1E99)` `▪left-rt=-1E-11` `GRAPH RANGE ZOOM TRACE SOLVE`
Repeat this process for $t = 20, t = 24, t = 28,$ and $t = 30$.	`exp=5.833333A/(1-.93…` `exp=520` `▪A=67.071127719429` `T=20` `bound=(-1E99,1E99)` `▪left-rt=0` `GRAPH RANGE ZOOM TRACE SOLVE` — `exp=5.833333A/(1-.93…` `exp=520` `▪A=78.160104385238` `T=30` `bound=(-1E99,1E99)` `▪left-rt=0` `GRAPH RANGE ZOOM TRACE SOLVE`
Make a table of the values of t and A as you find them. You need to find as many points as it takes for you to see the pattern the points are indicating when you plot them on a piece of paper.	

Step 3:   You need a model to draw a contour graph using your calculator. Even though there are several models that seem to fit the data points obtained in Step 2, their use would be misleading since the real best-fit model can only be determined by substituting the appropriate values in a multivariable function. The focus of this section is to use a graph to study the relationships between input variables, not to find the equation of a model for a contour curve. Thus, we always sketch the contours on paper, not with your calculator.

9.3 Partial Rates of Change

When you hold all but one of the input variables constant, you are actually looking at a function of one input variable. Therefore, all the techniques we previously discussed can be used. In particular, the calculator's numerical derivative **der1** can be used to find partial rates of change at specific values of the varying input variable. Although your calculator does not give formulas for derivatives, you can use it as discussed in Sections 4.4.1 and/or 4.4.2 of this *Guide* to check your algebraic formula for a partial derivative.

9.3.1 NUMERICALLY CHECKING PARTIAL DERIVATIVE FORMULAS

The basic concept behind checking your partial derivative formula is that your formula and the calculator's formula computed with **der1** should have the same outputs when each is evaluated at several different randomly-chosen inputs.

You can use the methods in Section 9.1.3 of this *Guide* to evaluate each formula at several different inputs and determine if the same numerical values are obtained from each formula. If so, your formula is *probably* correct.

You may find it more convenient to use your calculator's **TABLE**. If so, you must remember that when using the table, your calculator considers **x** as the variable that is changing. When finding a partial derivative formula, all other variables are held constant except the one that is changing. So, simply call the changing variable **x** and proceed as in Chapter 4.

Suppose you have a multivariable function $A(P, t) = P(1.061363551)^t$ and find, using derivative formulas, that $\frac{\partial A}{\partial t} = P (\ln 1.061363551) (1.061363551)^t$. We now illustrate how to check to see if you obtained the correct formula for $\frac{\partial A}{\partial t}$.

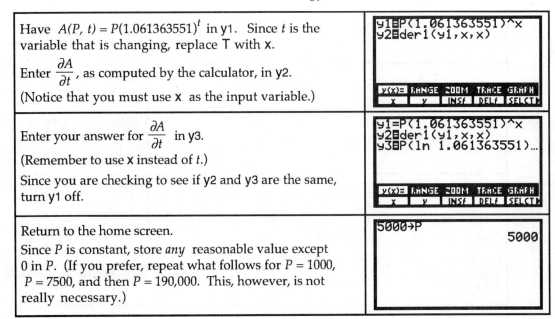

Have $A(P, t) = P(1.061363551)^t$ in y1. Since t is the variable that is changing, replace T with x. Enter $\frac{\partial A}{\partial t}$, as computed by the calculator, in y2. (Notice that you must use **x** as the input variable.)	y1⬛P(1.061363551)^x y2⬛der1(y1,x,x) y(x)= RANGE ZOOM TRACE GRAPH x ⎮ y ⎮ INSf ⎮ DELf ⎮ SELCT
Enter your answer for $\frac{\partial A}{\partial t}$ in y3. (Remember to use **x** instead of t.) Since you are checking to see if y2 and y3 are the same, turn y1 off.	y1=P(1.061363551)^x y2⬛der1(y1,x,x) y3⬛P(ln 1.061363551)… y(x)= RANGE ZOOM TRACE GRAPH x ⎮ y ⎮ INSf ⎮ DELf ⎮ SELCT
Return to the home screen. Since P is constant, store *any* reasonable value except 0 in P. (If you prefer, repeat what follows for $P = 1000$, $P = 7500$, and then $P = 190,000$. This, however, is not really necessary.)	5000→P 5000

| We now use the TABLE. Choose **ASK** and remember that with **ASK**, it does not matter what the Tblmin and ΔTbl settings are.

Input at least 4 values for $t = x$. | ```Press right arrow to scroll through table. ...335.43823) (2,335... ...425.66627) (6,425... ...,540.16435) (10,540... ...,685.46075) (14,685... ...,869.83979) (18,869...```
TI-85 | (table)
TI-86 |

TI-86 table:

x	y2	y3
2	335.4382	335.4382
4	377.8687	377.8687
8	479.5099	479.5099
10	540.1643	540.1643
25	1319.74	1319.74

x=

TBLST SELCT x y

Since the values shown for y2 and y3 are approximately the same, our formula for $\frac{\partial A}{\partial t}$ is *probably* correct.

- **TI-85** Note: Program **TABLE** will only allow you to **ASK** for values one at a time. If you prefer, you can choose **AUTO** with a ΔTbl that will encompass a variety of values for x. This was done to generate the TI-85 table above.

- Note that if you have rounded a model fit by the calculator before computing a partial derivative, then the y2 and y3 columns will very likely be *slightly* different.

- To check your formula for $\frac{\partial A}{\partial P}$, repeat the above procedure, but this time replace P by x and store any reasonable value (except 0) in T before going to the table. Remember that the variable that is changing always appears in the denominator of the partial derivative symbol.

9.3.2 VISUALIZING AND ESTIMATING PARTIAL RATES OF CHANGE

A partial rate of change of a multivariable function (evaluated at a specific point) is the slope of the line tangent to a cross-sectional model at a given location. We illustrate this concept with the two elevation cross-sectional models $E(0.8, n)$ and $E(e, 0.6)$.

Have $E(0.8, n)$ in y1 and $E(e, 0.6)$ in y2. Turn y1 on and y2 off using [F5] (SELCT). (**TI-86**: Also make sure all Stat Plots are off.)	```y1=-2.49999999991x^2... y2=-10.124103599x^3+...``` y(x)= RANGE ZOOM TRACE GRAPH x y INSf DELf SELCT
Press [EXIT] [F2] (RANGE or WIND). The values should be set to those shown to the right. Press [2nd] [EXIT] (QUIT) to return to the home screen.	```RANGE xMin=-.15 xMax=1.65 xScl=0 yMin=794.635 yMax=803.51 yScl=0``` y(x)= RANGE ZOOM TRACE GRAPH
Draw the tangent line to $E(0.8, n)$ at $n = 0.6$ with [2nd] [CUSTOM] (CATALOG) [–] (T), and use [▼] to select TanLn. Press [ENTER] to copy the instruction to the home screen, and press [2nd] [ALPHA] y [1] [,] .6 [)] [ENTER].	 y(x)= RANGE ZOOM TRACE GRAPH

Estimate the partial rate of change $\frac{\partial E(0.8,n)}{\partial n}$ at $n = 0.6$ by first pressing MORE F1 (MATH) F4 (dy/dx) and then pressing ◄ to trace to the approximate point of intersection. (Your view will probably be set in such a way that you are not able to locate $n = \text{x} = 0.6$ exactly.)	 1 x=.60714285714 y=800.08422119
Next, determine the slope of the tangent line at the point whose coordinates are shown at the bottom of the graphics screen in the previous step with ENTER. Remember that this is only an *estimate* of the true value of $\frac{\partial E(0.8,n)}{\partial n}$ at $n = 0.6$.	 dy/dx=-.5386554625
Now, turn y1 off and y2 on. Return to the home screen.	y1=-2.49999999991x^2... y2=-10.124103599x^3+... Y(X)= RANGE ZOOM TRACE GRAPH x y INSf DELf SELCT▶
Draw the tangent line to $E(e, 0.6)$ at $e = 0.8$ by pressing 2nd ENTER to return the last command, TanLn(y1, .6), to the home screen and use ◄ to edit the instruction to TanLn(y2, .8). Press ENTER.	
Estimate the partial rate of change $\frac{\partial E(e,0.6)}{\partial e}$ at $e = 0.8$ by first pressing MORE F1 (MATH) F4 (dy/dx) and then pressing ► to trace to the approximate point of intersection. (Your view will probably be set in such a way that you cannot locate $e = \text{x} = 0.8$ exactly.)	 2 x=.80714285714 y=800.11533519
Next, determine the slope of the tangent line at the point whose coordinates are shown at the bottom of the graphics screen in the previous step with ENTER. Remember that this is only an *estimate* of the true value of $\frac{\partial E(e,0.6)}{\partial e}$ at $e = 0.8$.	 dy/dx=.70114577316

9.3.3 FINDING PARTIAL RATES OF CHANGE USING CROSS-SECTIONAL MODELS   The procedure described above gave estimates of the partial rates of change. You will obtain a more accurate answer if you use your calculator's numerical derivative on the home screen to evaluate partial rates of change at specific input values.

Have *E(0.8, n)* in **y1** and *E(e, 0.6)* in **y2**.	`y1█-2.49999999991x^2…`
	`y2█-10.124103599x^3+…`
	
	`y(x)█ RANGE ZOOM TRACE GRAPH`
	`  x  │  y  │INSf│DELf│SELCT▶`

On the home screen, enter the instruction shown to the right to find $\frac{\partial E(0.8,n)}{\partial n}$ evaluated at $e = 0.8$, $n = 0.6$.
 	
(We are using **X** as the input variable because the cross-sectional models are in the **y=** list in terms of **x**.)
	`der1(y1,x,.6)`
	`              -.502949117647`
	
	`evalF│ nDer │ der1 │ der2 │ fnInt ▶`

On the home screen, enter the second instruction shown to the right to find $\frac{\partial E(e,0.6)}{\partial e}$ evaluated at $e = 0.8$, $n = 0.6$.	`der1(y1,x,.6)`
	`              -.502949117647`
	`der1(y2,x,.8)`
	`               .745328074181`

📖 9.4 Compensating for Change

As you have just seen, your calculator finds numerical values of partial derivatives using its **der1** function. This technique can also be very beneficial and help you eliminate many potential calculation mistakes when you find the rate of change of one input variable with respect to another input variable (that is, the slope of the tangent line) at a point on a contour curve.

9.4.1 EVALUATING PARTIAL DERIVATIVES OF MULTIVARIABLE FUNCTIONS

The previous section of this *Guide* indicated how to estimate and evaluate partial derivatives using cross-sectional models. Your calculator can be very helpful when evaluating partial derivatives calculated directly from multivariable function formulas. The most important thing to remember is that you must supply the name of the input variable that is changing and the values at which the partial derivative is evaluated.

Consider, for instance, the formula for a person's body-mass index:

$$B(h,w) = \frac{0.45w}{0.00064516h^2}$$

where h is the person's height in inches and w is the person's weight in pounds. We first find the partial derivatives $\frac{\partial B}{\partial h}$ and $\frac{\partial B}{\partial w}$ at a specific height and weight and then use those values in the next section of this *Guide* to find the value of the derivative $\frac{dw}{dh}$ at that particular height and weight.

Suppose the person in this example is 5 feet 7 inches tall and weighs 129 pounds.

In the previous section, we had to use **x** as the input variable because we were finding and graphing cross-sectional models. However, here we are given the multivariable formula.
 	
Enter *B(h, w)* in **y1** using the same letters that appear in the formula above.
	`y1█.45W/(.00064516H²)`
	
	`y(x)█ RANGE ZOOM TRACE GRAPH`
	`  x  │  y  │INSf│DELf│SELCT▶`

Let's first find $\dfrac{\partial B}{\partial h}$ at $(h, w) = (67, 129)$. Because h is changing, w is constant. Return to the home screen and store the value of the input that is held constant.
 	
Find the value of $\dfrac{\partial B}{\partial h}$ at $h = 67$ and $w = 129$ by evaluating the expression shown to the right.
 	
(Store this value for later use as N.)
	``
	129→W
	                    129
	der1(y1,H,67)
	        -.598329448472
	Ans→N
	        -.598329448472
	``
	`evalF  nDer  der1  der2  fnInt`
	
	H is the variable, and W is constant.

We now find $\dfrac{\partial B}{\partial w}$ at $(h, w) = (67, 129)$. Because w is changing, h is constant. Return to the home screen and store the value of the input that is held constant.
 	
Find the value of $\dfrac{\partial B}{\partial w}$ at $h = 67$ and $w = 129$ by evaluating the expression shown to the right.
 	
(Store this value for later use as D.)
	``
	67→H
	                     67
	der1(y1,W,129)
	         .155380128092
	Ans→D
	         .155380128092
	``
	`evalF  nDer  der1  der2  fnInt`
	
	W is the variable, and H is constant.

9.4.2 COMPENSATING FOR CHANGE

When one input of a two-variable multivariable function changes by a small amount, the value of the function is no longer the same as it was before the change. The methods illustrated below show how to determine the amount by which the other input must change so that the output of the function remains at the value it was before any changes were made.

We use the body-mass index formula of the previous section of this *Guide* to illustrate the procedure for calculating the compensating amount.

We first determine $\dfrac{dw}{dh}$ at the point $(67, 129)$ on the contour curve corresponding to the person's current body-mass index. The formula is $\dfrac{dw}{dh} = \dfrac{-\partial B}{\partial h} \div \dfrac{\partial B}{\partial w}$. An easy way to remember this formula is that whatever variable is in the numerator of the derivative (in this case, w) is the same variable that appears in the denominator of the partial derivative in the denominator of the quotient. This is why we stored $\dfrac{\partial B}{\partial w}$ as D (for denominator). The partial derivative with respect to the other variable (here, h) appears in the numerator of the quotient. Don't forget to put a minus sign in front of the numerator.

In the previous section, we stored $\dfrac{\partial B}{\partial h}$ as N and $\dfrac{\partial B}{\partial w}$ as D. So, $\dfrac{dw}{dh} = \dfrac{-\partial B}{\partial h} \div \dfrac{\partial B}{\partial w} = $ -N ÷ D. The rate of change is about 3.85 pounds per inch.	``` -N/D 3.85074626866 ```

To estimate the weight change needed to compensate for growths of 0.5 inch, 1 inch, and 2 inches if the person's body-mass index is to remain constant, you need to find
 	
$$\Delta w \approx \frac{dw}{dh}(\Delta h)$$
 	
at the various values of Δh.
	``
	              3.85074626866
	-N/D*.5
	              1.92537313433
	-N/D*1
	              3.85074626866
	-N/D*2
	              7.70149253731
	``

Chapter 10 Analyzing Multivariable Change: Optimization

10.2 Multivariable Optimization

As you might expect, multivariable optimization techniques on your calculator are very similar to those that were discussed in Chapter 5. The basic difference is that the algebra required to get the expression that comes from solving a system of equations in several unknowns down to one equation in one unknown is sometimes fairly difficult. However, once your equation is of that form, all the optimization procedures are the same as previously discussed.

10.2.1 FINDING CRITICAL POINTS To find critical points for a multivariable function (points at which maxima, minima, or saddle points occur), find the point or points at which the partial derivatives with respect to each of the input variables are zero. Once you use derivative formulas to find the partial derivative with respect to each input variable, set these partial derivatives formulas each equal to 0, and use algebraic methods to obtain one equation in one unknown input. Then, you can use the your calculator's solve routine to obtain the solution to that equation.

We illustrate these ideas with a model for the total daily intake of organic matter required by a beef cow grazing the Northern Great Plains rangeland:

$$I(s, m) = 8.61967 - 1.244s + 0.0897s^2 - 0.20988m + 0.035947m^2 + 0.214915sm \text{ kg per day}$$

Find the two partial derivatives and set each of them equal to 0 to obtain

$$\frac{\partial I}{\partial s} = {}^-1.244 + 0.1794s + 0.214915m = 0 \text{ and } \frac{\partial I}{\partial m} = {}^-0.20988 + 0.071894m + 0.214915s = 0$$

Solve one of the equations for one of the variables, say $\frac{\partial I}{\partial s}$ for s, to obtain

$$s = \frac{1.244 - 0.214915m}{0.1794}$$

(Remember that to solve for a quantity means that it must be by itself on one side of the equation and the other side of the equation cannot contain that letter at all.) Now, let your calculator do the rest of the work.

Warning: Everything that you do in your calculator depends on the partial derivative formulas that you have found using derivative rules. Double check your analytic work before you use any of the methods illustrated below.

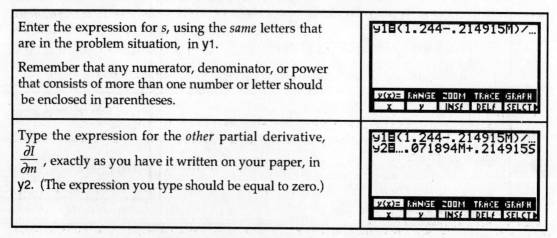

Enter the expression for s, using the *same* letters that are in the problem situation, in y1.
 	
Remember that any numerator, denominator, or power that consists of more than one number or letter should be enclosed in parentheses.
	`Y1B(1.244-.214915M)/...`
	
	`Y(X)E RANGE ZOOM TRACE GRAPH`
	`X Y INSF DELF SELCT`

Type the expression for the *other* partial derivative, $\frac{\partial I}{\partial m}$, exactly as you have it written on your paper, in y2. (The expression you type should be equal to zero.)	`Y1B(1.244-.214915M)/...`
	`Y2B....071894M+.214915S`
	
	`Y(X)E RANGE ZOOM TRACE GRAPH`
	`X Y INSF DELF SELCT`

Now, put the cursor in the y3 location. Type Rcl y2 and press ENTER . [Access with 2nd STO▶ (Rcl).]
 	
Next, replace *every* S in y3 with y1. What you have just done is substitute the expression for s in the other partial derivative!
	`y1B(1.244-.214915M)/…`
	`y2B-.20988+.071894M+…`
	`y3B…71894M+.214915y1`
	
	`y(x)= RANGE ZOOM TRACE GRAPH`
	`   x   y   INSf  DELf  SELCT▶`

The only thing that is left is to solve the equation y3 = 0. Use the solve routine to do this.
 	
You can try several different guesses and see that they all result in the same solution. The best approach, however, is to note that y3 is a linear equation and therefore has only one root.
	`exp=y3`
	`  exp=0`
	`▪M=6.8998840512348`
	`  bound=(-1E99,1E99)`
	`▪left-rt=-2E-14`
	
	`GRAPH RANGE ZOOM TRACE SOLVE`

Return to the home screen and type M to recall the value you just found. Remember that you have the expression for s in y1, so just type y1 to find the value of s. The calculator, however, doesn't know this is s, so store the answer in memory location s.
 	
Note that if you now find the value of y2, it should be 0.
	`                6.89988405123`
	`y1`
	`              -1.33159744075`
	`Ans→S`
	`              -1.33159744075`
	`y2`
	`                       2E-14`

Either type the multivariable function I(s, m) in one of the unused locations in the y= list and evaluate it or just type I(s, m) on the home screen. (You already have the values you found for s and m stored in those variables.)
 	
What all this has done is locate a critical point of I(s, m) at s ≈ ⁻1.33 and m ≈ 6.90.
	`8.61967-1.244S+.0897S`
	`2-.20988M+.035947M²+.`
	`214915S*M`
	`              8.72384977581`

- It is **very important** to note that because the TI-85 and the TI-86 accept strings of letters as variable names, whenever you perform an operation on two separate variables, you must put the indicated symbol for that multiplication. For instance, we typed the product of S and M in the above evaluation as S*M. This is necessary; if it were not done, the calculator would think we meant a single variable called "SM".

10.2.2 THE DETERMINANT TEST

Now that you have found one or more critical points, the next step is to classify those point(s) as points at which a maximum, a minimum, or a saddle point occurs. The Determinant Test will often give you the answer. Because this test uses derivatives, the calculator's numerical derivative der1 can help you.

For the beef cattle example in Section 10.2.1 with critical point s ≈ ⁻1.33 and m ≈ 6.90, you now need to calculate some second partial derivatives evaluated at this point. Recall that we have $\frac{\partial I}{\partial m} = I_m$ in y2. However, y1 does not contain $I_s = \frac{\partial I}{\partial s}$.

Clear y1 and enter $\frac{\partial I}{\partial s}$ = ⁻1.244 + 0.1794s + 0.214915m .
 	
(We will not use y3 again, so you can also delete that location if you wish.)
	`y1B-1.244+.1794S+.21…`
	`y2B-.20988+.071894M+…`
	
	`y(x)= RANGE ZOOM TRACE GRAPH`
	`   x   y   INSf  DELf  SELCT▶`

Return to the home screen, and recall S and M to be certain that the critical point values are still stored in those locations. If not, store the critical values in those locations. **This is a very important step!**
 	
(You will have better accuracy with the Determinant Test if you use several decimal places for the values.)
	``
	M
	            6.89988405123
	S
	           -1.33159744075
	``

Now, if we take the derivative of y1 = I_s with respect to s, we will have I_{ss} , the second partial derivative of $I(s, m)$.   Find $I_{ss} \approx 0.1794$.
 	
If we take the derivative of y2 = I_m with respect to s, we will have I_{ms} , the partial derivative of $I(s, m)$ with respect to m and then s.   Find $I_{ms} \approx 0.2149$.
	``
	der1(y1,S,-1.331597)
	                  .1794
	der1(y2,S,-1.331597)
	                .214915
	``
	`evalF  nDer  der1  der2  fnInt`
	
	Be certain the value you type after the variable matches the input that is varying.

Now, if we take the derivative of y2 = I_m with respect to m, we will have I_{mm}, the second partial derivative of $I(s, m)$.   Find $I_{mm} \approx 0.0719$.
 	
If we take the derivative of y1 = I_s with respect to m, we will have I_{sm}, the partial derivative of $I(s, m)$ with respect to s and then m.   Find $I_{sm} \approx 0.2149$.
	``
	der1(y2,M,6.8998841)
	                .071894
	der1(y1,M,6.8998841)
	                .214915
	``
	`evalF  nDer  der1  der2  fnInt`
	
	Be certain the value you type after the variable matches the input that is varying.

Now, find the value of $D \approx (0.1794)(0.0719) - (0.2149)^2$. Since $D < 0$, the Determinant Test tells us that the point $(s, m, I) \approx (^-1.33, 6.90, 8.72)$ a saddle point.	```
	.1794*.071894-.214915
	2
	          -.033290673625
	``
	`evalF  nDer  der1  der2  fnInt`

- The values of the second partial derivatives were not very difficult to determine without the calculator in this example. However, with a more complicated function, we strongly suggest using the above methods to provide a check on your analytic work to avoid making simple mistakes.

- Instead of the sequence of operations that was illustrated above, you could enter the multivariable function in y1 and use the calculator's second derivative, der2, to take the second partials without having to find the first partial derivatives!

📖 10.3 Optimization Under Constraints

Optimization techniques on your calculator when a constraint is involved are exactly the same as those discussed in Sections 10.2.1 and Section 10.2.2 except that there is one more equation involved in the analytic process.

10.3.1 CLASSIFYING OPTIMAL POINTS UNDER CONSTRAINED OPTIMIZATION

Use the methods indicated in Section 10.2.1 to find the critical point $L = 7.35$, $K = 39.2$. We illustrate the procedure for determining if this is the point at which a maximum or minimum occurs for the Cobb-Douglas Production function $f(L, K) = 48.1L^{0.6}K^{0.4}$ subject to the constraint $g(L, K) = 8L + K = 98$.

Find $f(L, K)$ at $L = 7.35$, $K = 39.2$.
 	
 We now use the methods of Section 10.2.2 to test close points that are on the constraint.
	```7.35→L:39.2→K:48.1L^.6*K^.4```
	       690.608479798
	
	 `evalF  nDer  der1  der2  fnInt ▸`

We now test *close* points. You need to remember that whatever close points you test, they must be near the critical point and *they must be on the constraint* $g(L,K)$. Be wary of rounding during the following procedure. Rounding of intermediate calculations and/or critical points can give a false result when the "close" point is very close to the optimal point.

Choose a new value of L that is less than L, say 7.3. Determine the value of K so that $8L + K = 98$:
 	
 $8(7.3) + K = 98$, so $K = 98 - 58.4 = 39.6$.
 	
 Now find $f(L, K)$ at $L = 7.3$, $K = 39.6$.
	```7.3→L:39.6→K:48.1L^.6*K^.4```
	       690.584564125
	
	 `evalF  nDer  der1  der2  fnInt ▸`

Choose another nearby value of L, but this time choose one that is more than L, say 7.4. Determine the value of K so that $8L + K = 98$:
 	
 $8(7.4) + K = 98$, so $K = 98 - 59.2 = 38.8$.
 	
 Now find $f(L, K)$ at $L = 7.4$, $K = 8.8$.
	```7.4→L:38.8→K:48.1L^.6*K^.4```
	       690.584455414
	
	 `evalF  nDer  der1  der2  fnInt ▸`

Since $f(7.35, 39.2) \approx 690.608$ is greater than both $f(7.3, 39.6) \approx 690.585$ and $f(7.4, 38.8) \approx 690.585$, we conclude that $L = 7.35$ thousand worker hours, $K = \$39.2$ thousand is the point at which the maximum value of $f(L, K)$ occurs. Rounding this answer to make sense in the context of the problem, we find that 691 mattresses is a maximum production level.

- When you test close points, it is best to use all the decimal places provided by the calculator for the optimal values when classifying optimal points since rounding will, in many cases, give inaccurate or misleading results in the classification.

Chapter 11 Dynamics of Change: Differential Equations and Proportionality

📖 11.1 Differential Equations and Accumulation Functions

Many of the differential equations we encounter have solutions that can be found by determining an antiderivative of the given rate-of-change function. So, many of the techniques we learned with the calculator's numerical integration function apply to this chapter. (See Chapters 6 and 7 of this *Guide*.)

11.1.1 EULER'S METHOD FOR $dy/dx = g(x)$

You may encounter some differential equations that cannot be solved by standard methods. You may want to draw an accumulation graph for a differential equation without first finding an antiderivative. In either of these cases, Euler's method might help you. Euler's method relies on the use of the derivative of a function to approximate the change in the function. Recall from Section 5.3 of *Calculus Concepts* that the approximate change in *f* is the rate of change of *f* times a small change in *x*. That is,

$$f(x + h) - f(x) \approx f'(x) \cdot h \text{ where represents the small change in } x.$$

Now, if we let $b = x + h$, and $x = a$, the above expression becomes

$$f(b) - f(a) \approx + f'(a) \cdot (b - a) \text{ or } f(b) \approx f(a) + (b - a) \cdot f'(a)$$

The first values we choose for the coordinates of the point (a, b) are given to you and are often called the initial condition. We then use the formula given above to involve the slope of the tangent line at *a* to approximate the change in the function between *a* and *b*. When *h*, the distance between *a* and *b*, is fairly small, Euler's method will often give close numerical estimates of points on the solution to the differential equation containing $f'(x)$.

Be wary of the fact that there is some error involved in each step of the approximation process that is compounded when each result is used to obtain the next result.

We illustrate Euler's method with the differential equation giving the total sales, in billions of dollars, of a computer product:

$$\frac{dS}{dt} = \frac{6.544}{\ln(t + 1.2)} \text{ billion dollars per year}$$

where *t* is the number of years after the product is introduced.

Because Euler's method involves a repetitive process, a program that performs the calculations described above can save you time and avoid possible errors in calculation.	`y1=6.544/ln (x+1.2)`
	
	 `y(x)= RANGE ZOOM TRACE GRAPH`
	 `  x    y    INSf  DELf SELCT►`

To use this program, you must have the differential equation in location y1 of the y(x)= list with x as the input variable.	

Add program EULER to your list of programs. (The program is found in the TI-85/86 Appendix.)	`EULER`
	
	 `NAMES EXIT`
	 `  DIFF  EULER LOGIS LSLIN MDVAL►`

Run the program. Each time the program stops for input or for you to view a result, press ENTER to continue.
 	
We are told in this example to use 16 steps. Enter this value. We are also told to use steps of size 0.25. Enter this value.
	`HAVE dy/dx in y1`
	
	`Number of steps= 16`
	`Step size= .25`

The initial condition is given as the point (1, 52.3). Enter these values when prompted for them.	`HAVE dy/dx in y1`
	
	`Number of steps= 16`
	`Step size= .25`
	`Initial input= 1`
	`Initial output= 53.2`

The first application of the formula gives an us an estimate for the value of the quantity function at $x = 1.25$:
 	
$S(1.25) \approx 55.275$.
	`Number of steps= 16`
	`Step size= .25`
	`Initial input= 1`
	`Initial output= 53.2`
	`Input,Output is`
	`                1.25`
	`     55.2749378245`

Continue pressing ENTER to obtain more estimates of points on the quantity function S.	`        71.7020961191`
	`Input,Output is`
	`                4.75`
	`     72.6420741685`
	`Input,Output is`
	`                   5`
	`     73.5594275692`

When the number of steps has been completed, the program draws a graph of the estimates connected with straight line segments. This is an approximation to the graph of the solution of the differential equation.	

11.1.2 EULER'S METHOD FOR $dy/dx = h(x, y)$ Program EULER can be used when the differential equation is a function of x and y with $y = f(x)$. Follow the same process as illustrated in Section 11.1.1 of this *Guide*, but enter $\frac{dy}{dx}$ in y1 in terms of both x and y.

If the differential equation is written in terms of variables other than x and y, let the derivative symbol be your guide as to which variable is the input and which is the output. For instance, if the rate of change of a quantity is given by $\frac{dP}{dn} = 1.346P \cdot (1 - n^2)$, you would enter y1= 1.346y(1 − x)², using 2nd ALPHA 0 (y) to type y.

When the differential equation is given in terms of x and y, such as $\frac{dy}{dx} = 5.9x - 3.2y$, enter y1= 5.9x − 3.2y. The differential equation may be given in terms of y only. For instance, if $\frac{dy}{dx} = k(30 - y)$ where k is a constant, enter y1= K(30 − y). Of course, you need to store a value for k or substitute a value for k in the equation before using the program. It is always better to store the exact value for the constant instead of using a rounded value.

TI-85/TI-86 Calculator Appendix

Programs listed below are referenced in *Part B* of this *Guide*. They should be transferred to you via a cable using the LINK mode of a TI-85 or a TI-86, transferred to your calculator using the TI-GRAPH LINK™ cable and software for a PC or Macintosh computer and a disk containing these programs, or, as a last resort, typed in your calculator. Refer to your owner's *Guidebook* for instructions on typing in the programs or transferring them via a cable from another calculator.

(Instructors who have the TI-GRAPH LINK™ software can contact the author of this *Guide* at ibbrh@clemson.edu and request that the programs be sent to them via e-mail or on a computer disk. Be sure to specify whether you use a Macintosh or a PC-compatible computer for either method of obtaining the programs. All programs can then easily be transferred to students.)

The programs and the chapter of *Calculus Concepts* in which each program is first referenced are listed below. The programs given below run on both the TI-85 and TI-86 graphing calculators, but not all the programs are needed for the TI-86. All these programs can be transferred directly from a TI-85 to a TI-86 and vice-versa.

PROGRAM NAME	PROGRAM SIZE (bytes)	CHAPTER FIRST REFERENCED	CALCULATOR
DIFF	889	1	TI-85, TI-86
STPLT	380	1	TI-85
TABLE	485	1	TI-85
LOGISTIC	2069	2	TI-85
MDVAL	171	6	TI-85
NUMINTGL	1514	6	TI-85, TI-86
SINREG	1406	8	TI-85
EULER	303	11	TI-85, TI-86

OPTIONAL

LSLINE	614	1	TI-85, TI-86
SECTAN	581	4	TI-85, TI-86

The code for each of the programs follows. If you have to type in these programs rather than having them transferred from another calculator or a computer, it is strongly suggested that you compare the line-by-line instructions given in the code with what you type in your calculator. Both the TI-85 and the TI-86 distinguish between upper and lower case letters, so be sure you use the same case as in the program code. Even one misplaced symbol or letter will cause the program to not properly execute.

```
DIFF              • Program
 :ClLCD
 :Disp "Store x-values in L1"
 :Disp "Store y-values in L2"
 :Disp "Continue?"
 :Menu(1,"Yes",GO,2,"Quit",NO)
 :Lbl NO
 :Stop
 :Lbl GO
 :dimL L1→M
 :dimL L2→N
 :If M≠N:Goto Z
 :N-1→dimL L6
 :For(H,1,N-1,1)
 :L1(H+1)-L1(H)→L6(H)
 :End
 :For(H,1,N-2,1)
 :If L6(H+1)≠L6(H)
 :Goto FN
 :End
 :N-1→dimL L3
 :For(A,1,N-1,1)
 :L2(A+1)-L2(A)→L3(A)
 :End
 :N-2→dimL L4
 :For(B,1,N-2,1)
 :L3(B+1)-L3(B)→L4(B)
 :End:ClLCD
 :Disp "Choice?"
 :Lbl RC
 :Menu(1,"1st",FST,2,"2nd",SEC,3,
     "%",PER,4,"Quit",QT)
 :Lbl FST
 :Disp "1st differences in L3"
 :Disp L3
 :Goto RC
 :Lbl SEC
 :Disp "2nd differences in L4"
 :Disp L4
 :Goto RC
 :Lbl PER
 :N-1→dimL L5
 :1→E
 :For(E,1,N,1)
 :If L2(E)==0:Goto W
 :End
 :For(E,1,N-1,1)
 :(L3(E)/L2(E))*100→L5(E)
 :End
 :Disp "Percent change in L5"
 :Disp L5
 :Goto RC
 :Lbl Z
 :Disp "Lists are of unequal"
 :Disp "length. Check data."
 :Stop
 :Lbl W
 :0→dimL L5
 :Disp "Percent change not"
```

```
(Program DIFF continued)
 :Disp "calculated...cannot"
 :Disp "divide by 0"
 :Goto RC
 :Lbl QT
 :Stop
 :Lbl FN
 :Disp "Input values not"
 :Disp "evenly spaced"
 :Stop
```

```
EULER              • Program
 :ClLCD
 :0→dimL L1:0→dimL L2
 :FnOff
 :Disp "HAVE dy/dx in y1"
 :Disp ""
 :Input "Number of steps= ",N
 :Input "Step size= ",H
 :Input "Initial input=   ",x
 :Input "Initial output= ",y
 :For(I,1,N,1)
 :x→L1(I)
 :y→L2(I)
 :y1→T
 :x+H→x
 :y+H*T→y
 :Disp "Input,Output is"
 :Disp x
 :Disp y
 :Pause
 :End
 :x→L1(N+1):y→L2(N+1)
 :STPLT
 :xyline (L1,L2)
```

```
LOGISTIC            • Program
:ClLCD
:Disp "Data in L1,L2"
:Disp ""
:Disp "Enter continues"
:Pause
:ClLCD
:If dimL L1≠dimL L2:Then
:Disp "List lengths not"
:Disp "equal"
:Stop
:End
:dimL L1→M
:M→dimL L5
:max(L2)*1.01→Z
:Z→C
:sum L1→T
:ln L2→L6
:sum L6→V
:0→N
:Repeat (C≤Z) or (N==3)
:ln (-1*L2+C)→L5
:sum L5→U
:(-sum (L5*L1)+sum (L6*L1)+(U-
   V)/M*T)/(sum (L1²)-T²/M)→B
:e^((U-V+B*T)/M)→A
:e^(-B*L1)→L4
:(1+A*L4)⁻¹→L5
:sum (L2*L5)/sum (L5²)→C
:End
:C*L5→L3
:sum ((L3-L2)²)→E
:0→N
:√.5ᴇ-10→Y
:2*Y→R
:ClLCD
:Outpt(2,1,"STEP:")
:Outpt(6,1,"SSE:")
:While (R>Y) and (N<20)
:N+1→N
:Outpt(2,7,N)
:Outpt(6,6,E)
:(3,1)→dim MatA
:C→MatA(1,1):A→MatA(2,1):B→
   MatA(3,1)
:Outpt(3,2,"Working...")
:e^(-B*L1)→L4
:(1+A*L4)⁻¹→L5
:C*L5→L3
:L4*L5²→L6
:(3,1)→dim MatC: -2*sum ((L2-
   L3)*L5)→MatC(1,1)
:2*C*sum ((L2-L3)*L6)→MatC(2,1)
:-2*C*A*sum ((L2-L3)*L6*L1)→
   MatC(3,1)
:(3,3)→dim MatE
:2*sum L5²→MatE(1,1)
:2*sum ((L2-2*L3)*L6)→MatE(1,2)
:MatE(1,2)→MatE(2,1)
```

(Program LOGISTIC continued)

```
:2*A*sum ((2*L3-L2)*L6*L1)→
   MatE(1,3)
:MatE(1,3)→MatE(3,1)
:2*C*sum ((C*L6-2*(L2-L3)*L4*
   L5)*L6)→MatE(2,2)
:2*C*sum ((-C*A*L6+(2*A*L5*L4-
   1)*(L2-L3))*L6*L1)→MatE(2,3)
:MatE(2,3)→MatE(3,2)
:2*C*A*sum ((C*A*L6-(2*A*L5*L4-
   1)*(L2-L3))*L6*L1²)→MatE(3,3)
:Outpt(3,2,"Computing ")
:-1*MatE⁻¹*MatC→MatD
:2*E→F
:10→S
:-1→I
:While (F>E) and (S>0)
:I+1→I
:If I>5:Then
:0→S
:Else
:Outpt(4,2,"Computing ")
:Outpt(4,12,I)
:.1*S→S
:S*MatD+MatA→MatB
:MatB(1,1)→C
:MatB(2,1)→A
:MatB(3,1)→B
:e^(-B*L1)→L4
:C*(1+A*L4)⁻¹→L3
:sum (L3-L2)²→F
:End
:End
:Outpt(4,2,"              ")
:If S==0:Then
:Outpt(3,2,"Still working")
:-1*MatC→MatD
:(MatCᵀ*MatC)*(MatCᵀ*MatE*MatC)⁻¹
   →MatB
:10*MatB(1,1)→S
:-1→I
:While (F>E) and (S>0)
:I+1→I
:If I>5:Then
:0→S
:Else
:Outpt(4,2,"Computing ")
:Outpt(4,11,I)
:.1*S→S
:S*MatD+MatA→MatB
:MatB(1,1)→C
:MatB(2,1)→A
:MatB(3,1)→B
:e^(-B*L1)→L4
:C*(1+A*L4)⁻¹→L3
:sum (L3-L2)²→F
:End
:End
```

(Program LOGISTIC continued)

```
:Outpt(4,2,"                   ")
:If S==0:Then
:Outpt(3,2,"No improvement")
:Else
:Outpt(3,2,"                   ")
:End
:End
:F→E
:{0,0}→L6
:S*MatD(2,1)*MatA(2,1)⁻¹→L6(1)
:S*MatD(3,1)*MatA(3,1)⁻¹→L6(2)
:abs L6→L6
:max(L6)→R
:End
:{R,Y}
:y1=L/(1+A*e^(-B x))
:C→L
:ClLCD
:Disp "The model is"
:Disp "Y=L/(1+A*e^(-B x))"
:Outpt(4,2,"L=")
:Outpt(4,4,L)
:Outpt(6,2,"A=")
:Outpt(6,4,A)
:Outpt(7,2,"B=")
:Outpt(7,4,B)
```

LSLINE   • Program

```
:FnOff
:0→A:0→B:1→C
:y1=A+B x
:L1→xStat
:L2→yStat
:dimL xStat→N
:ClLCD
:Disp "You will next view"
:Disp "the data. Use tick"
:Disp "marks to guess the"
:Disp "slope and y-intercept"
:Disp "of best fit line."
:Disp "xScl=":Outpt(6,7,xScl)
:Disp "yScl=":Outpt(7,7,yScl)
:Pause
:Lbl A1
:Scatter
:Pause
:Disp ""
:Input "slope=",B
:Input "y intercept=",A
:1→K:0→x:0→S
:Lbl V
:xStat(K)→x
:y1→Y
:(yStat(K)-Y)²+S→S
:Line(xStat(K),yStat(K),x,Y)
:K+1→K
:If K≤N
:Goto V
:Pause
:Disp "SSE=",S
:Pause
:If C==2
:Goto W
:Input "Try again? Y(1) N(2) ",C
:If C==1
:Goto A1
:LinR
:ShwSt
:Pause
:DrawF RegEq
:Pause
:a→A:b→B
:1→K:0→x:0→S
:Goto V
:Lbl W
:FnOff
```

```
MDVAL              • Program
:Input "Function location? ",H
:dimL L1→N
:N→dimL L2
:For(j,1,N,1)
:L1(j)→x
:If H==1:y1→L2(j)
:If H==2:y2→L2(j)
:If H==3:y3→L2(j)
:If H==4:y4→L2(j)
:End
:L2


NUMINTGL           • Program
:ClLCD
:Disp "Enter f(x) in y1"
:Disp ""
:Disp "Continue?"
:Menu(1,"Yes",YS,2,"No",NO)
:Lbl NO:Stop
:Lbl YS
:Disp ""
:Disp "Draw Pictures?"
:Menu(1,"Yes",YE,2,"No",NR)
:Lbl YE:1→H:Goto LE
:Lbl NR:2→H
:Lbl LE
:ClLCD
:Input "Left endpoint? ",A
:Input "Right endpoint? ",B
:If H==1:Then
:A→xMin:B→xMax
:iPart ((B-A)/20)→W
:If W==0:0.1→W
:seq(V,V,A,B,W)→L5
:dimL L5→N:N→dimL L6
:For(j,1,N,1)
:L5(j)→x:y1→L6(j)
:End
:real L6→L6
:min(L6)→yMin
:If yMin>0:0→yMin
:max(L6)→yMax
:If yMax<0:0→yMax
:W→xScl
:iPart ((yMax-yMin)/10)→yScl
:0→dimL L5:0→dimL L6
:End
:Lbl A0
:ClLCD
:Disp "Enter Choice:"
:Disp "Left Rect  (1)"
:Disp "Right Rect (2)"
:Disp "Trapezoids (3)"
:Input "Midpt Rect (4) ",R
:Lbl A1
```

(Program NUMINTGL continued)
```
:ClDrw
:Disp ""
:Input "N? ",N
:(B-A)/N→W
:0→S:1→C
:Lbl A2
:If R==1:Goto A3
:If R==2:Goto A4
:If R==3:Goto A3
:If R==4:Goto A5
:Lbl A3
:A+(C-1)W→x
:x→J:x+W→L
:Goto A7
:Lbl A4
:A+C*W→x
:x-W→J:x→L
:Goto A7
:Lbl A5
:If H≠1:Then
:If N>5:Then
:1→Z:W/2→H:A→x
:Lbl A8
:x+H→x:y1+S→S
:A+Z*W→x
:IS>(Z,N):Goto A8
:S*W→S:Goto T
:End:End
:A+C*W-W/2→x
:x-W/2→J
:x+W/2→L
:Goto A7
:A→G:G+W→G:G→V
:Lbl A9
:V→x:y1→Y:V+W→x
:4Y+2y1+S→S
:V+2*W→V
:If V<B:Goto A9
:G-W→x:y1→E
:B→x:y1→F
:(W/3)*(S+E-F)→S
:Goto T
:Lbl A7
:y1→K:K+S→S
:If H==1:Goto D
:Lbl I
:IS>(C,N)
:Goto A2
:If R==3:Then
:A→x:y1→P
:B→x:y1→Q
:S+(Q-P)/2→S
:End
:W*S→S
:Lbl T
:Disp "SUM=",S
:Pause
```

(Program NUMINTGL continued)

```
:ClLCD
:Lbl E
:ClLCD
:Disp "Enter Choice:"
:Disp "Change N        (1)"
:Disp "Change Method (2)"
:Input "Quit           (3) ",T
:If T==1:Goto A1
:If T==2:Goto A0
:If T==3:Goto F
:Lbl F
:Stop
:Lbl D
:If R==3:Then
:x→I:L→x
:y1→M:I→x
:Else:K→M
:End
:Line(J,0,J,K)
:Line(J,K,L,M)
:Line(L,M,L,0)
:If C==N:Pause
:Goto I
```

SECTAN   ● Program

```
:ClLCD
:ClDrw:2→R
:Disp ""
:Disp "Have f(x) in y1 and"
:Disp "draw graph of f"
:Disp ""
:Disp "Continue? "
:Menu(1,"Yes",YS,2,"NO",NO)
:Lbl YS
:Disp ""
:Disp "x-value of point"
:Input "of tangency? ",A
:Lbl RT
:Disp ""
:Disp "Press ENTER to "
:Disp "see secant lines"
:If R==1:Goto RS
:Disp "from the left"
:Goto RU
:Lbl RS
:ClDrw
:Disp "from the right"
:Lbl RU
:Disp "approach tangent"
:Disp "line"
:Pause
:(xMax-xMin)/3→K
:If K>50:48→K
:For(J,1,5,1)
:A-K→x
:If R==1:A+K→x:x→D
:y1→B:A→x:y1→C
:(B-C)/(D-A)→M
:A→x:y1→E
:DrawF (M(x-A)+E)
:K/2→K
:End
:Pause
:If R==1:Goto RV
:1→R:Goto RT
:Lbl RV
:ClLCD
:Disp "Press ENTER to"
:Disp "see tangent line"
:Pause
:ClDrw
:TanLn(y1,A)
:Lbl NO:Stop
```

```
SINREG              * Program
 :ClLCD
 :Disp "Have data in L₁,L₂"
 :Disp ""
 :Disp "Enter continues"
 :Pause
 :ClLCD
 :dimL L1→N
 :If N≠dimL L2:Then
 :Disp "Lists are of unequal"
 :Disp "length.  Check data."
 :Stop
 :End
 :0→P
 :Repeat P>0
 :Input "Period Guess? ",P
 :End
 :Disp "Maximum"
 :Input "Iteration? ",M
 :max(min(M,16),1)→M
 :0→A
 :2π/P→B
 :0→C
 :OneVar L2:x̄→D
 :(10^-10)*sum (L2-D)²→P
 :13→dimL L3
 :For(I,1,M)
 :Fill(0,L3)
 :For(K,1,N)
 :{B*L1(K)}→L4
 :{cos L4(1),sin L4(1)}→L4
 :{L4(1),L4(2),L1(K)*(C*L4(1)-
    A*L4(2)),A*L4(1)+C*L4(2)+
    D-L2(K)}→L4
 :{L4(1)²,L4(1)*L4(2),L4(1),L4(
    1)*L4(3),L4(1)*L4(4),L4(2)²,
    L4(2),L4(2)*L4(3),L4(2)*L4(4)
    ,L4(3),L4(4),L4(3)²,L4(3)*L4(
    4)}+L3→L3
 :End
 :[[L3(1),L3(2),L3(3),L3(4),L3(
    5)][L3(2),L3(6),L3(7),L3(8),
    L3(9)][L3(3),L3(7),N,L3(10)
    ,L3(11)][L3(4),L3(8),L3(10),
    L3(12),L3(13)]]→MatA
 :For(K,1,4)
 :For(J,K+1,4)
 :If abs MatA(J,K)>abs MatA(K,K)
 :Then
 :rSwap(MatA,K,J)
 :End:End
 :If MatA(K,K)≠0:Then
 :multR(1/MatA(K,K),MatA,K)→MatA
 :For(J,1,4)
 :If J≠K
 :mRAdd(-MatA(J,K),MatA,K,J)→MatA
 :End:End:End
 :Disp I
 :If I>1 and MatA(4,4)==0:Then
 :M→I
 :Else
```

(Program SINREG continued)
```
 :A-MatA(1,5)→A
 :B-MatA(4,5)→B
 :C-MatA(2,5)→C
 :D-MatA(3,5)→D
 :L3(5)*MatA(1,5)+L3(9)*MatA(2,5)
 :L3(11)*MatA(3,5)+L3(13)*MatA
    (4,5)→K
 :If K<P:M→I
 :End:End
 :A→P
 :√(P²+C²)→A
 :If P<0:Then
 :-.5*π-tan⁻¹ (C/P)→C
 :Else
 :.5*π-tan⁻¹ (C/P)→C
 :End
 :y1=A*sin (B x+C)+D
 :0→dimL L3
 :0→dimL L4
 :ClLCD
 :Disp "The model is"
 :Disp "Y=A sin (B x+C)+D"
 :Outpt(4,2,"A=")
 :Outpt(4,4,A)
 :Outpt(5,2,"B=")
 :Outpt(5,4,B)
 :Outpt(6,2,"C=")
 :Outpt(6,4,C)
 :Outpt(7,2,"D=")
 :Outpt(7,4,D)
```

```
STPLT               * Program
 :dimL L1→N
 :min(L1)→A1
 :max(L1)→B1
 :min(L2)→C1
 :max(L2)→D1
 :(B1-A1)/10→X1:(D1-C1)/10→Y1
 :A1-X1→xMin:B1+X1→xMax
 :C1-1.5*Y1→yMin:D1+Y1→yMax
 :0→xScl
 :0→yScl
 :(xMax-xMin)/125.5→P1
 :(xMax-xMin)/128.5→P2
 :(yMax-yMin)/62→P3
 :For(K,1,N,1)
 :L1(K)-P2→W1
 :L1(K)+P1→W2
 :L2(K)-P3→H1
 :L2(K)+P3→H2
 :Line(W1,H2,W2,H2)
 :Line(W2,H2,W2,H1)
 :Line(W2,H1,W1,H1)
 :Line(W1,H1,W1,H2)
 :End
```

TABLE   • Program
```
:ClLCD
:Disp "Have function(s) in"
:Disp "y(x)= list."
:Func
:1→y
:Disp "Enter 1 (AUTO) or"
:Input "2 (ASK) ",W
:If W==2:Then
:Prompt x
:x→A:x→B:1→H
:Goto J:End
:Input "TblMin? ",A
:Input "∆Tbl? ",H
:Input "Number of values? ",V
:(V-1)*H+A→B
:Lbl J
:For(x,A,B,H)
:eval x→N
:dimL N→D
:{y,D}→dim Z
:For(C,1,D)
:(x,N(C))→Z(y,C)
:End
:y+1→y
:End
:ClLCD
:Disp "Press ENTER to see"
:Disp "[(input, output)]."
:Disp "1st column is for y1,"
:Disp "2nd column for y2,..."
:Disp " "
:Disp "Press right arrow to "
:Disp "scroll through table."
:Pause
:" "
:round(Z,5)→Z
:Z
```